Every Drop
for Sale

Jeffrey Rothfeder

Jeremy P. Tarcher/Penguin
a member of Penguin Group (USA) Inc.

New York

Every Drop for Sale

Our Desperate Battle Over Water in a World About to Run Out

Most Tarcher/Penguin books are available at special quantity discounts
for bulk purchase for sales promotions, premiums, fund-raising, and
educational needs. Special books or book excerpts also can be created
to fit specific needs. For details, write Penguin Group (USA) Inc.
Special Markets, 375 Hudson Street, New York, NY 10014.

Jeremy P. Tarcher/Penguin
a member of
Penguin Group (USA) Inc.
375 Hudson Street
New York, NY 10014
www.penguin.com

The Library of Congress cataloged the hardcover edition as follows:

Rothfeder, Jeffrey, date.
Every drop for sale : our desperate battle over water in a world
about to run out / by Jeffrey Rothfeder.
p.   cm.
Includes index.
ISBN 1-58542-114-6
1. Water-supply—Economic aspects.   I. Title.
HD1691.R67   2001   2001027903
333.91—dc21

ISBN 1-58542-367-X (paperback edition)

Printed in the United States of America
1   3   5   7   9   10   8   6   4   2

Book design by Meighan Cavanaugh

To my kids, Alexis and Ben.

May your children be born into a world that

finally respects the Earth and protects water, its greatest gift.

Acknowledgments

THIS BOOK COULD not have been completed without the significant help of Robert Ratner, a professor of English and journalism at Florida International University. He was my writing partner twenty-two years ago, when I first became a journalist. We reported and wrote stories for *The Washington Post* and *The St. Petersburg Times* on expensive and harmful southern U.S. water projects with wildly exaggerated promises of benefits. That was an exciting time to be a journalist. There was no Internet, no instant information; if you wanted the facts, you had to ask probing questions, dig through paper documents, and be in Washington, D.C., one day and Mobile, Alabama, the next to speak to newly uncovered sources and look people in the eye to detect whether they were telling the truth or not. Those journalistic methods are the only ones I know, and they were the ones I employed in researching and writing this book—albeit with some help from the Web and electronic communications.

Robert Ratner was a big part of the process. He conducted research, pored over material with me during hours of conversations, provided voluminous notes to use as needed, and continually pushed me to fine-tune my themes, questioning my logic at crucial moments and guiding me when I was unclear. He is among the best reporters I've met, and he made sure that this book reached the standards of the highest principles of journalism.

My editor, Wendy Hubbert, also played a vital role. The idea to do a book about water was hers in the first place. While our concepts for the book initially didn't dovetail—she wanted something more evocative, and I was looking for something more hard-hitting—we agreed on enough to take a chance. A good decision, as it turns out. Under Wendy's guidance and patience, the book straddled our two sensibilities without compromising either of us. Just as important, Wendy has a deft touch as an editor, leaving the writer's words mostly intact while tweaking the organization and presentation of the story perfectly, to make what might initially seem dull come to life.

The biggest joy I had in writing this book was in meeting or speaking to many remarkable people who gave me and my researchers their time and tolerated my naiveté as they shared their thoughts about the water crisis. These include environmental experts such as Maude Barlow, Shirley Solomon, Dan Sullivan, and Elden Hughes; water specialists such as Douglas MacDonald, Woody Wodraska, and Ron Gastelum; scientists such as Jay Searles, Deborah Kelley, Penny Boston, and Jeffrey Karson; and individuals including Oscar Olivera and Claro Rodriguez, who in their day-to-day lives face the consequences of what we have done to the world's water supply.

Two other people deserve special mention. Jim Shultz, a reporter for the Pacific News Service who covered the Cochabamba water wars in Bolivia with passion and tireless on-the-ground work, is an example for all young journalists. And Jane Clement, my friend who shared stories of desperate water scarcity in Africa, has devoted a key part of her life to creating a sustainable environment for the people of Kenya. She is now beginning to turn her hard efforts into reality, and I am convinced that her organization, the Foundation for World Health and Development (www.fwhd.org), will be highly successful, if only because Jane's heart is so deep and her force of will so powerful.

Other friends of mine were extremely supportive, and I owe them a debt of gratitude. They listened, sometimes for long stretches, to the accounts of water scarcity and water mismanagement I came across in my research, and they encouraged me by their reactions and their endless fascination for the theme and narratives of the book. My appreciation to John McCorry, Rebecca Cox, Jim Bartimo, Ann Graham, Mary Behr, Marianne Mustafa, Curtis Lang, and my agent, John Ware.

There are two people I must thank above all others: Jim and Teena Lockhart, the residents of Rapid City, South Dakota, whom I met soon after the devastating 1972 flood there. They inspired me with their love for each other and for their children; their integrity in trying to help their neighbors during the disaster—many of whom perished; their fatalistic wisdom and humor in the aftermath of the nightmare; their love for their environment; and their sincere understanding of the apparently capricious and sometimes violent nature of water. It was only after I met them—and especially after I shared many conversations with Teena and sensed her keen understanding of the nature of life on the planet and our place in it—that I realized how meaningless existence is when it is lived without respect for the world we have been given. I am glad to have had the chance to tell the Lockharts' story. Rest in peace, Teena.

Contents

Preface to the 2004 Edition *xii*

Beginning: Scenes of a Crisis *1*

1. Controlled Fury 15

2. Invisible Desert 43

3. Right or Need? 77

4. Private Ties 99

5. To the Highest Bidder 119

6. The Scramble to Restore 139

7. There Are No Winners 153

8. Planet Water 173

Ending: Salvation *187*

Notes *193*

Index *201*

Preface to the 2004 Edition

SOME THINGS HAVE changed since *Every Drop for Sale* was first published, in October 2001. Little things, meaningless things. One example: I wrote that Enron, through its subsidiary Azurix, was positioning itself to become a major player in international water sales. In other words, Enron's experience in marketing oil and gas would be transferred to water. Obviously, with Enron bankrupt and Azurix now a part of American Water Works, that didn't happen.

But, sadly, important aspects of the global water crisis haven't changed at all. As when the book was released, one-third of the world's population must sustain itself without enough water—fifty liters per person per day—for a minimum quality of life. And people are still dying in huge numbers—well over 10,000 a day—from waterborne diseases. Given our heightened level of interest when a few hundred people die from SARS, a half dozen die from bird flu, and a couple hundred die from mad cow disease over a decade, it's hard to comprehend why daily, avoidable water deaths are neglected and why the world's water woes are barely on the radar of policymakers, media, and even residents of developed countries. In the poor nations of Africa, Asia, and Central and South America, where populations suffer the most from shortages of clean water, people don't have the luxury of ignoring the subject.

I also reported in the book the growing number of water conflicts around the world, which are destined to play out in wealthy as well as poor countries. The pitched arguments over dwindling water supplies in the Middle East and parts of Africa continue unabated, and they've worsened in the United States. Fights over who has the right to the water in the Colorado River and the Ogallala Aquifer have gotten so tense that the Department of the Interior recently issued a list of "hot" water-depleted spots in the U.S. that, according to the agency, must be addressed now if major water conflicts between neighboring states and cities are to be avoided. Among the locations to be dealt with are Las Vegas and Reno, Nevada; Albuquerque, New Mexico; Denver, Colorado; Houston, Texas; Salt Lake City, Utah; and Flagstaff, Arizona.

Privatization of water supplies, too, still plagues both developed and developing countries. One striking example is South Africa, unique among nations for including in its post-apartheid constitution the right of citizens to clean water. Now, only about one-sixth of the population faces water shortages, while under apartheid about one-third of the population lacked access to clean drinking water. Privatizing water supplies—that is, turning the delivery of water over to private corporations—is central to the South African government's water policy but, in some cases, this has produced disastrous results. In a few communities, after privatization the price of water rose so quickly that local residents couldn't afford the clean water they were being offered. Bills went unpaid. Eventually, the companies administering the water supplies turned the taps off or required people to prepay for water that they would use. Before long, poorer residents were forced to go back to polluted lakes, rivers, and streams to get water. A spiral of disease and deaths ensued.

Despite South Africa's experience, I still believe that privatization is necessary to supply ample water for everybody in the world. Many local governments—even in the United States—simply do not have the resources to manage and maintain heavily burdened water collection and distribution systems. But privatization works only when the contract setting up the water management agreement includes a series of price controls and possible subsidies for poor people. Also, the privatization effort must be overseen by powerful and mindful regulatory authorities, with citizen participation.

Other solutions to the water crisis—particularly technological ones like desalinization of ocean water—have also not progressed very much since the book was published.

I had hoped that three years after the original publication of *Every Drop for Sale* a great many strides in appreciating and respecting water as a precious, shared resource and as an individual right—not just a need—would have been made. I also had hoped that the water crisis would convincingly demonstrate that responding to global issues—be they environmental or geopolitical—requires much more than local solutions, and that intractable international matters cannot be resolved when the protection of parochial interests is the sole motivation. So far, none of this has occurred—a disturbing testimony to how relevant *Every Drop for Sale* remains.

Beginning: Scenes of a Crisis

There went up a mist from the Earth,
and watered the whole face of the ground.

GENESIS 2:6

THE CHATTAHOOCHEE RIVER rises harmlessly enough. At its source, caressing the low slopes of the Appalachian Mountains in northern Georgia, it's only a trickle of water—a thin, barely visible serpentine streak dampening the hills, giving them the brown color and smooth texture of turned pottery. In that setting, it looks more like a leak in the bedrock than a headwaters, so insignificant it barely catches your eye.

But as the Chattahoochee skirts south, consumes tributaries, widens into a stream, and then turns into a full-fledged river, it quickly loses its innocence. Even before completing its 130-mile stretch from the Appalachians to Atlanta, the Chattahoochee becomes the object of a nasty tug-of-war. Three states—millions of people—are fighting over who owns the river's water. It seems so improbable, while canoeing its rapids or sitting by its waterfalls in the rural highland valley, but this dispute—and everything surrounding it—is killing the Chattahoochee.

In Georgia's view, the Chattahoochee is its domain. The river is the sole water supply for Atlanta's metropolitan area, which has come to be the embodiment of worst-case urban sprawl. The city's uncontrolled population boom—from 2.2 million in 1980 to 3.7 million in 2000—has forced it to stretch into its surrounding areas, building tract homes, housing developments, shopping malls, office parks, and industrial centers.

Providing water for this expansion has severely taxed the Chattahoochee. To put the problem in perspective: when Atlanta's largest treatment plant opened in 1991, it tapped 3.8 billion gallons of the river's water each year; now it pumps nearly 20 billion gallons annually. Moreover, the increase in demand occurs as the Chattahoochee is becoming more and more polluted, mostly the result of runoff from new roofs that are being constructed and asphalt in the river's watershed as soil gives way to concrete. If, as expected, Atlanta's population reaches 5 million by 2025, the Chattahoochee won't be able to handle the load. The river will be bone dry, at least as a water source, and the city will run out of freshwater.

That specter isn't slowing down Atlanta's growth. Instead, the city is aggressively making plans to squeeze even more water out of the Chattahoochee by building as many as a dozen additional dams and reservoirs on the river and its tributaries. This, in turn, has raised the ire of Alabama and Florida, the two downstream states that rely on the Chattahoochee for water as well. Both claim that Georgia is stealing the river for itself, hoarding its waters and depleting the supply available for them. And farmers in southern Georgia are siding with Alabama and Florida against Atlanta, as their irrigation allotment is threatened.

This water war has created a pitched level of distrust in the Southeast, as everybody tries to gain control of as many last drops of water as they can get. John "Bubba" Johnson, who farms five hundred acres on the Flint River (which runs into the Chattahoochee at the Florida border), impressed upon me how deep the suspicions are running and how critical this water fight is becoming as we walked near a field of alfalfa. He twisted his work boot into the ground, turning his toe left and right as if he were crushing a bug. Then he lifted his shoe.

"Touch the boot," he said. "It's dry."

It was, caked only with a loamy, pitted soil; there was no moisture. His land always used to be wet, Johnson said, as water freely seeped upward from the Flint's aquifers. A few years ago, just a bare amount of pressure from his boot would have created a puddle of water.

"If I lose my water, I'm out of business," he said. "That's how important it is. We have a right to what God has put down here for us."

Locked as they are in the middle of this struggle—and blinded by their own need for water—everyone involved in this dispute is missing the point. This isn't just a regional issue or an anomalous little argument between neighbors. The American Southeast is just one of many places that have failed to adequately manage their water supply and that are now running out of usable water. Just as the Chattahoochee is drying up because it can't keep pace with increasing demand, the pipelines that support Los Angeles and the other desert towns in the Southwest are emptier as well. Meanwhile, in midwestern cities, where most of the great rivers have been channeled and dammed, pollution and floods make the search for freshwater more critical every day.

In the United States, the water crisis is perceived as a series of local problems, serious enough but resolvable. And while we may accurately use the phrase "running out of water" to describe the dilemma, it's difficult for even the most pessimistic of us to envision the prospect of literally not having enough water to drink or take a bath. But in other countries, that's exactly what this phrase means.

In Vietnam, for instance, only 50 percent of urban residents and 25 percent in the countryside have access to water in their homes—and that water is often not sanitary. The water purification plants and pipeline systems throughout the country are archaic, built by the French during the first part of the twentieth century. In many regions, exposed-pit latrines are the only means for carrying waste from the houses to the sea.

And in areas where no water pipes exist, it's not uncommon to see open pools dug behind homes to catch the rain. Residents use this water untreated for cooking, drinking, and washing. Some hospitals are reusing water left in basins and sinks for more than one patient, because there isn't enough freshwater to go around. Most jarring, the percentage of Vietnam's population with access to safe drinking water is dropping precipitously—to about 36 percent in the mid-1990s, from 45 percent ten years earlier.

To counter this crisis, Vietnam's water chief has put together a plan that will by 2020 provide on average 120 to 150 liters of safe water per day to each resident of Vietnam, more than three times what they receive today. But even if the country is able to raise the tens of billions of dollars the plan will cost, the average Vietnamese citizen would still receive just a bit above the absolute daily minimum amount of water needed for survival. The country would remain far, far behind the United States, for example, where the availability of more than enough water is barely ever doubted—even in regions where it should be—and San Pellegrino sipped casually during a meal equals one person's total daily water allotment in Cambodia and Mozambique.

Peter Gleick, one of the cofounders of the Pacific Institute for Studies in Development, Environment and Security, has developed a fascinating formula for determining how much water people need each day to survive. While most minimum daily water targets set by organizations like the World Bank and the World Health Organization consider only drinking and sanitation, Gleick's numbers include two other basic needs: bathing and cooking. Taking all four into account, Gleick came up with a figure of 50 liters per day (lpd) as the amount of water each individual requires to meet a minimum quality of life. The breakdown is 5 lpd for drinking, 10 lpd for cooking, 15 lpd for bathing and 20 lpd for sanitation.

Put in context, these numbers are remarkable. Most toilets in the United States—that is, those built before the mid-1990s—flush 23 liters down their pipes with each use. That's about half Gleick's daily benchmark. At the same time, nearly 2.2 billion people spread out among sixty-two countries—one-third of the world's population—live below the minimum water level. Some quite a bit below. The worst conditions are in Haiti and Gambia, where people subsist on an average of 3 liters of water per day. Imagine having less than two large bottles of Poland Spring as your entire water ration for the day. More developed countries are also on the list. In India, only 31 lpd per person is available for domestic consumption, and families pay 25 percent of their income for drinking water; in Kenya, 36 lpd; in Bolivia, 41 lpd; and in the Dominican Republic, 48 lpd.

Even regions that have an abundance of water face similarly dire situations. Lima, Peru, for instance, gets a plentiful supply of water from the Amazon as well as runoff from the Andes, yet its residents pay up to two dollars—in many cases, more than half of their daily salary—for 500 liters of often-contaminated water. By contrast, each resident of the United States and Canada uses at least that amount of clean water per day, and the percentage of their salaries used to pay for it is so negligible that most people in those countries don't know how much water costs them.

The lack of usable water in a large portion of the world exacts an enormous toll. Upward of 10 million deaths per year, mostly among the young and elderly, are caused by water-related diseases, chiefly cholera and dysentery. Nearly 250 million new cases are reported annually. The leading cause of infant death worldwide is unsafe water. This huge gap between water haves and have-nots has actually grown in the past decade; in 1990, 1.1 billion people—half the current number—in fifty-five countries fell below the minimum.

"It's a human tragedy, which could explode into a human bloodbath," Gleick says.

Initially I had a difficult time grasping the importance of his statistics. I had become too inured to global tragedy, as many of us in developed nations are. Extreme and endless poverty, natural disasters that seem unimaginable, inexplicable wars between ancient tribes—all of these resulting in millions of deaths a year in undeveloped countries—are endemic or occur with regularity in hundreds of nations. But they're barely talked about anymore in the rest of the world. A combination of elements makes them even less newsworthy than the dog-bites-man story: they're (take your pick) too commonplace, too incomprehensible, too alien, too far away, or too heinous to report.

It's the same with Gleick's grim analysis of water scarcity: it's disturbing but not surprising that a large number of people are dying of thirst and living in unhygienic squalor. It's easy for people in the developed nations to dismiss Gleick's prediction of a coming human bloodbath over environmental conditions—over water starvation—as something that will probably take place a fair distance away and thus won't affect us very much, if it happens at all.

Then something unforeseen occurred. One hot, muggy evening in summer, during my second shower of the day, as limitless water poured over me, I couldn't escape the disquieting observation that in just a few moments under the cooling spray, I used up the entire daily water ration of a person in Mali or Somalia. Pregnant, diseased, or just hot and dirty, residents of those countries got a measly 10 liters per day at most, Gleick had said; I consumed that much water under the shower in the time it took to think about the inequity. That was a turning point for me. I could no longer avoid putting a human face—or more accurately, many human faces—to Gleick's statistics. And what made this realization more disquieting was that it occurred again and again at inconsequential moments when water was usually not even on my mind: doing the third load of wash, turning on the dishwasher, running a faucet for nearly a minute to warm the water before brushing my teeth at night. Such trivial moments, to me, turn out to be the essence of life to someone someplace else.

By quantifying the water problem with such an unambiguous benchmark, Gleick made an impact on me that all the dense position papers and hand-wringing of water policy experts never could. The idea that 50 liters per day is a dividing line between an individual's survival and slow, spiraling demise was such a tangible and troubling expression of the water crisis that it haunted me.

That's how this book took shape. I decided to investigate and experience the world's water crisis—all sides of it—and to explore how serious it actually is; in other words, to do firsthand what I was already doing in my mind: put faces to the grim statistics. I set out to examine the world's hot spots that have the least amount of water, as well as the places that have so much of it that plans are in the works to sell the excess to the highest bidder, wherever the highest bidder may be. I wanted to find out which ideas about water management have failed us and left us with such widespread inequity and at the brink of apparent disaster. I hoped to unearth solutions—to meet people who had the power to do something about the water dilemma and who actually were taking steps to fix it. And I wanted to answer the most difficult question: What happens if we don't resolve the crisis?

I talked to people in Europe, Asia, Latin America, Africa, and the United States, some in developing countries, some in high places, and some so far in the back woods that they have no reliable communication systems. As different as these regions are, the similarities of their feeble water management strategies—and their inability to learn from mistakes and then to try different, saner approaches—was striking. It isn't only the poorest countries, I found, that are increasingly desperate to devise ways to deliver water to their residents. Some of the richest regions are also panicking. In the past, they used their wealth to paper over water shortages. Now, with water scarcity growing, they can't do that as easily anymore, and places like southern California have become so frantic about finding more water that they're digging dry holes in the desert hoping to uncover a gushing aquifer underneath. In my research for this book, I met people who were willing to die for 50 liters per day, and others who controlled water and wanted to charge so much for it that, indeed, dying might be the only way for those who need it to get it. And I visited the scarred survivors of the bizarre water practices we've instituted—the attempts to manage and control water that actually accomplished just the opposite, producing any number of disasters ranging from deadly floods to pollution to carcinogen-carrying dust storms.

I discovered one disturbing fact almost immediately, a fact that's probably the primary reason for the mess we're in. Water may be the single most critical element of life—it is the largest and most complex habitat on Earth, nourishes every species on the planet, and is the primary material of all living things—but we don't have a clue about its true nature. To be sure, we have scientific knowledge of it, but that hasn't helped, because water is too unpredictable. It defies simple rules. Unlike anything else on the planet, water doesn't bend to brute force and machines, the tools we've otherwise used to tame our environment. Water reacts differently, its own way, from one situation to another. It's almost emotional and emotionless in its responses—a combination that we're not particularly equipped to handle. Consequently, we've been powerless to mold it to fit the structure of our political, social, and environmental systems, so much so that it may as well have come from Mars (which could actually be the case).

At the root of our confusion about water is the sheer amount of it that exists on Earth. We'd like to believe that the supply of it is infinite, and at first glance it seems to be. But the truth is, even though our planet is 70 percent water, we're running out of freshwater. Actually, there wasn't very much to start with: only 2.5 percent of the water on Earth is not salt water. And with much of that frozen in northern glaciers, humans have access to less than 0.08 of 1 percent of the total water on the planet.

But even that amount is dwindling, and not only because of thirsty people. Individual humans use less than 10 percent of our planet's freshwater. Much more, from 60 to 70 percent, goes to irrigation, and the rest is taken up by industry. With population growth and expanding economies, demands for water are skyrocketing, even taking into account conservation measures. In fact, the 6 billion people on Earth today—projected to grow to 8 billion by 2025—share the same amount of water that was available to less than one-sixth of this population at the turn of the nineteenth century. And while dozens of studies predict what future water requirements will be, the consensus is that the total amount of water needed for people, for producing food for the swelling population, and for industry will increase by as much as 45 percent in the next twenty years.

Freshwater is renewable only by rainfall, which produces 40 to 50 cubic kilometers per year. The High Plains Ogallala Aquifer, which runs thirteen hundred miles from Texas to South Dakota, is being used up eight times faster than nature can refill it. In Mexico City, pumping exceeds net recharge by as much as 50 to 80 percent a year. In Israel extraction exceeded replacement by 2.5 billion cubic meters in the last twenty-five years. In Africa the aquifers barely refill at all—they're being depleted by 10 billion cubic meters per year. The water table under Beijing dropped 37 meters in the last forty years. The land under Bangkok has actually sunk due to overpumping.

With Earth's ecosystems bundled so tightly together, destruction of forestland is at least partly to blame for the depletion of aquifers. Under the best conditions, the vast ground cover of forests filters and seeps water slowly underground, where it can be stored for people to use. In the continental United States, for instance, national forests, which account for only 8 percent of the nation's acreage, are responsible for nearly 15 percent of aquifer replenishment. But sprawl and commercial development have eaten up forestland—as much as 50 percent of it around the world since the beginning of history, according to the World Resources Institute, an environmental think tank. And as the forests have dwindled, there's less soil and earth to stop rainfall from driving directly toward rivers without stocking the water table first.

What's more, our rivers, streams, and lakes—essentially our main water supply—have been so mismanaged by decades of damming, diversion, and industrial pollution that the amount of usable water they provide is falling rapidly. In the United States, only 2 percent of the rivers aren't dammed. Almost none of the water in the Colorado and the Rio Grande ever reaches the sea. With the completion of the Aswan High Dam in Egypt, the Nile stopped flowing freely, and much less of it now empties into the Mediterranean. The Ganges in India is in the same predicament. In England, 30 percent of the rivers are down to one-third their average depth. The Aral Sea, in the former Soviet Union, once the world's largest body of freshwater outside North America, has shrunk to half its previous size and is now little more than three salty lakes. Lake Chad in Central Africa, which once encompassed 10,000 square miles, has been pumped dry by excessive irrigation. It now covers only about a tenth of its former area. As less freshwater makes its way to the sea, the Earth's ability to replenish its water supply is reduced. That's because most water evaporation, the precursor to precipitation, takes place in the ocean.

Meanwhile, pollution of freshwater stocks is also diminishing the amount of available water. Seventy-five percent of Poland's rivers are so polluted, the water is unfit even for industrial use. Thirty-seven percent of America's freshwater fish are at risk of extinction, 51 percent of crayfish, 40 percent of amphibians, and 67 percent of freshwater mussels—mostly from poisoned water running off from agriculture.

In short, what looked like an endless supply of water when the first aqueducts, dams, and water diversion schemes were devised perhaps as many as eight thousand years ago—projects that continue to be built even now—has become an unbalanced supply-and-demand equation.

This widening inequity is already beginning to destabilize certain regions of the world. In fact, with water scarcity worsening, many people believe that if there is a World War III, it will be fought over water. Some countries are already nearly at war over it. Botswana and Namibia are arguing about who has rights to the Okavango River. Malaysia is threatening to cut off Singapore's water supply. Turkey opposes Kurdish independence in part because the Kurds live in the mountains where the water is. And Syria, fearful that the Turks could control its water supply if they gain control of the Kurdish region, has dropped its long-standing support of the Kurds and ingratiated itself with Turkey. Israel, Jordan, and other Middle Eastern countries are fighting over water—so much so that even after signing a peace treaty in 1994 with Israel, the late King Hussein of Jordan remarked that the only thing he could imagine the two nations going to war over was water.

With tension over water escalating in these and dozens of other places, awareness is growing that international security is vitally and sensitively linked to the environment. In response, national security cabals that are focused solely on water diplomacy have quietly been formed in almost every major country and at key worldwide economic organizations, such as the Agency for International Development. Their main purpose is to identify potential water-war hot spots and open diplomatic channels to stop disputes before they erupt into full-scale war.

People have always tried to manage water for political and economic gain, to keep it from hurting them in the case of floods, and to provide for their basic needs. But controlling water is a risky venture—one that ultimately almost always carries a price. In this instance, it was a steep one: we've slowly depleted the amount of water available for our use. This depletion took place even while our need for water increased—a need that was partially spurred, ironically, by the political and economic gains that water management helped provide. It occurred to me chillingly that for thousands of years we've been playing a zero-sum game without knowing it—and we've been losing.

That conclusion convinced me that the world's water crisis is at a serious stage, as region after region struggles to provide enough clean water for its residents, its agriculture, and its industry. And as the scramble for water worsens, there's been a significant new development in water's world order: the emergence of a strong private market where water is being bought and sold as a commodity. It was inevitable. With demand rising, supply dwindling, and traditional water management approaches producing more problems than solutions, the real value of water is increasing rapidly. Suddenly companies, like countries before them, see the potential for profits in water.

As a result, all kinds of new private ventures have arisen to move water from one place to another and to deliver water to people who don't have enough of it. Some companies are transporting it in huge, sausagelike bags pushed by tugboats, others in hulls of ships. Some are using old-fashioned pipelines and thereby hauling water over much longer distances than ever before; and still others are taking over water supply management from public utilities and imposing strict income-and-loss scenarios on pricing where there were none before. Two French firms, Vivendi SA and Suez Lyonnaise des Eaux, own or control water companies in 120 countries on five continents and distribute water to almost 100 million people—giving them the kind of buying power over water that a near monopoly like De Beers has over diamonds. In addition, speculators who see the prospects for huge paydays when worldwide water markets go into full swing are grabbing up water rights.

The water entrepreneurs remind me of the people who founded high-tech start-ups in the 1990s. Their language is hip corporate modern and full of New Age optimism. They're part of a fast-changing business with huge upside: water has the potential to be more profitable than oil. And being first is critical, while being ready as the business environment takes off is essential.

It's hard not to get caught up in their excitement—until you realize that what they're talking about is water, nature's most essential element, where life began. Looked at that way, the international water business is disquieting. In effect, the growth of the private sector thrusts water into the heart of globalization; it's just another item to be bought and sold—like oil, stocks, money, computer chips, or automobiles—in an economically interdependent world.

Think of the ramifications, though. When the market places a value on water—not as a natural element but as a commodity—economic market forces and not environmental considerations determine its future. And ultimately ours. It sounds suspiciously like what we've been doing, in one form or another, for five thousand years by putting development and controlling water for commercial purposes first, and the protection of water supplies second. Will it be another minus in the zero-sum game?

I met some people from Cochabamba, Bolivia, a city in the Andes, who warned me of the dark side of water globalization. The city turned its water system, admittedly not very efficient in the first place, over to Bechtel Corporation, a U.S. company. To pay for fixing up the infrastructure and modernizing the pipelines into homes and businesses, Bechtel raised water prices to a level higher than those in Washington, D.C. This in a city where most of the people have less money than the homeless in the United States. Riots broke out over the privatization of the water supply—"Dirty water is better than water we can't afford," one of the Cochabambinos told me. During the riots, the Bolivian police killed a seventeen-year-old boy in the streets, and thousands were injured. Bechtel gave up and left town, but it was a hollow victory. While dirty water may be better than water out of your price range, it's no substitute for clean water, which the Cochabambinos still don't have.

This incident reflects a fundamental change that water globalization has brought about. Embracing the philosophy of privatization, world trade and political leaders have replaced the concept of water as a human *right* with the concept that water is only a human *need*. It's a subtle difference semantically, but it carries with it a discomforting meaning: that water isn't a shared natural resource that must be provided to everyone, to the greatest extent possible with no compromises, but rather a commodity that people need but have no birthright to.

This shift in attitude—endorsed even by some environmentalists who have somehow adopted the skewed view that unless a price tag is placed on water, people won't value it enough to protect it—has only exacerbated the possibility that the struggle over the world's water could be a desperate one, more a free-for-all than a cooperative effort. The aims of its participants can be described simply: to control, through physical and political management, the world's limited water supply. Whatever countries, companies, or people achieve this control will command the lion's share of power, capital, and influence in the world, especially as the amount of usable water per person dwindles during this century.

Therein lies the full importance of Gleick's data. The fight over water, at its least complicated, is actually being waged to determine who will have enough water tomorrow and who will go thirsty—the boundary line in Gleick's statistics. The nations capable of delivering at least the minimum required amounts of water to their citizens will be the dominant survivors; they will influence and control countries that can't. Similarly, the most successful private companies will be those best able to profit from satisfying the growing water needs of those who already have enough, while providing badly needed water to those who don't.

This book takes a look at a world that is already in crisis over water, politically, environmentally, socially, and economically. But it's impossible to explore water without, in effect, examining who we are. Water is the only natural element that provides a reflection of ourselves; long before mirrors, it alone showed us what we looked like. It is doing that metaphorically now as well. The world's water crisis is forcing us to face our limitations and the limitations of nature. It's making us look at ourselves to see if we have the capacity to figure out how to share what nature has provided to sustain us, or if we can only fight over it.

These are human issues, as I found in researching this book, played out on political and economic stages. But people whose relationship to water is far less complicated face them more quietly every day, people whose lives moved me deeply: flood victims; women struggling to raise families in regions where rusted pipes drip a few drops of water once a day; odd coalitions of longtime ideological and territorial foes, now united to protect the river environments they share; people whose land was stolen out from under them for the water rights and are victims of the environmental disasters that ensued. Too often their stories get lost in the noise of the water wars, in the high-level geopolitical maneuvering and the scramble for profits. In this book, they're the foundation for the broader narrative. The reason for this approach is as logical to me as Gleick's data: humans can survive for weeks without food but only days without water. Our bodies are made up of about 65 percent water and our brains are 75 percent. Water is a right, not a need. And we forget too easily that what we're fighting over is nothing more than the simple action of having a drink of water.

Controlled Fury

*The hollow sea-shell . . . when held against the ear . . .
we hear the faint, far murmur of the breaking flood . . .
It is the blood
In our own veins.*

EUGENE LEE-HAMILTON, "SEA-SHELL MURMURS"

IT WAS SIMPLE to overlook the fact that Canyon Lake Dam even existed. About twelve feet above the waterline and five hundred feet across, in the high summer heat the leathery-faced earthen structure faded into the sun-baked background in Rapid City, South Dakota. At other times of the year, it resembled nothing more than a cracked wall pounded by bitter winds, a blot more than an attraction.

It wasn't all that useful anyway. It was just a recreational dam, one that had outlived its own life span; it didn't irrigate any farms or produce any hydroelectricity. Its only real purpose was to keep neglected Canyon Lake, which had been carved out of the dam's flooded back plain, from spilling into Rapid Creek.

When the dam was built in the late 1800s, the brochures said that the new lake it created would be an idyllic spot for Rapid City residents, perfect for fishing, boating, swimming, even wooing under the moon. Nearly a century later, there was little if any of that going on at Canyon Lake. Local residents found it convenient enough to drive to better-stocked fishing holes on the Missouri River, a few hundred miles to the east, or to haul a boat to one of the many big wide-crested lakes in Montana to the north-west, or to fly to a more picturesque and romantic spot on the Pacific Ocean.

But on June 9, 1972, Canyon Lake Dam finally put itself on the map. In the midst of a driving rainstorm, its tiny spillway filled with debris and sediment from the waterlogged lake bed. With the spillway clogged and water unable to flow freely anymore through Canyon Lake Dam, the searing pressure of the lake behind it and Rapid Creek rising in front of it sand-wiched the dam like a vise, eventually rending a three-hundred-foot breach in its muddy wall. Ripped apart, Canyon Lake Dam crumbled at its own feet, letting loose a torrent of water that shot out with the force of a bat-tering ram and engulfed all of Rapid City below it. Two hundred thirty-six people were killed, more than in any American flood in recent memory.

In the days that followed, as bits of earth and mud—the debris of Canyon Lake Dam—floated like body parts in the creek, more Rapid City residents trekked out to gawk at the misshapen structure than had ven-tured there in the entire decade or so before. They looked at their feet to the now-free-flowing, placid water in the creek. And then they glanced southward to the rubbled city, the site of the disaster, where homes and cars were piled every which way, wherever they had landed after being plucked from their moorings by the water, weightless like toy figurines. Some shook their heads in disbelief.

"I guess they should never have built this thing," said a seventy-year-old man as he rubbed his callused hands through his thin hair. His daugh-ter and son-in-law had died in the flood. "It never seemed to do us any good."

He said this absentmindedly, lost in his own mourning, to no one in particular. But nearly thirty years later, his words have taken on a signifi-cance that would have been ill mannered to even consider back then, standing in the well of the deadly disaster's aftermath.

Without even attempting to, and speaking more from his emotions than from his intellect, this man put his finger on arguably the fundamental explanation for the water crisis we face today. Throughout history, from the very earliest civilizations, people have built hundreds of thousands of dams and projects that piggyback on dams—abatements, levees, canals, navigational pools, and irrigation ditches—blindly and arbitrarily, without considering the consequences. We constructed them with one purpose: to alter for our own needs the natural course of water. Many of these water projects, as the old man observed, should never have been built, and many of them haven't really done us any good.

In one sense, we had no choice but to try to control water. It was almost an involuntary action. People, as we've learned clearly by now, are driven more than anything by market forces, by the desire to improve their lives with commerce, transactions, robust economies, and profits. As we have strived to fulfill these aspirations, water has been both an impediment and an opportunity. Unmanaged, it's a natural and geographical obstacle, flooding away whole communities at will, and in the process destroying carefully constructed economic systems, while making it difficult and expensive to move products and raw materials from one place to another because of its ubiquity on Earth. But water also offers a potential advantage: those who were wealthy or creative enough to control the water supplies or the irrigation channels or the barge canals—or perhaps those with the most and best weapons—produced societies with larger, healthier, better-employed, and more innovative populations, with more items to sell to more places around the world and more ways to get their products to buyers.

To control water, thus, has long seemed to be obviously the better option; all the benefits were on that side of the equation. The other course—not to build dams or otherwise restrict water's movement, and not to make water a partner in the search for profits and commerce—in other words, opting not to harness this powerful force—simply would have gone against our desire for progress and wealth.

Managing water has always been easy to justify even without having to admit that economic gain was the propelling reason behind it. Many high-minded goals and achievements could be offered. Irrigation, for instance, allows modern farming, greatly expanding the world's food supply. Flood control projects stop rivers from spreading into local communities and killing thousands of people each year. Reservoirs store water in wet periods for use in drier times by farmers, industry, and cities. Hydroelectric plants extend power to even the most rural areas everywhere in the world.

These were tangible and in some cases extremely worthwhile results. But in almost all instances the potential for environmental and social damage wasn't considered; it was overwhelmed by the passion for more and bigger water projects. The upshot has been an imposing record of water management activities that crisscross every region of the world and every civilization that has inhabited them. The first dams were likely built as long as eight thousand years ago in the valleys of the Tigris and Euphrates rivers. There, it is believed, ancient civilizations constructed sophisticated agriculture irrigation canals, which were filled by streams diverted by tiny dams of brushwood and earth.

The earliest unearthed remains of dams date from about 3000 B.C. and belong to the complex water system that served the town of Jawa, in what is now Jordan. The largest of these dams, which redirected water into a series of interlocking reservoirs with underground pipelines, was 13 feet high and 262 feet long. A few hundred years later, construction began on a dam on the Nile. Intended to be a massive 46-foot-high structure of sand, gravel, and rock retained by 17,000 blocks of cut stone, it was washed away before it could be completed. Wisely, it would be nearly five thousand years before the Egyptians would attempt to build another dam on the river.

Few societies, however, viewed the failure of the Nile dam and many other subsequent ones as a warning. Because water management frequently accomplished its main purpose of generating profits as well as commercial and political power, in the thousands of years since the first dams were constructed, efforts to control water became increasingly aggressive and increasingly irrational.

The first attempt to harness water purely to mollify megalomania may have occurred in twelfth-century Sri Lanka. At that time King Parakrama Bahu, who could only generously be termed a tyrant, went on a water management spree during which he claimed to be responsible for more than four thousand huge dams. One of them was a massive fifty feet tall and eight miles long, the biggest dam in the world by volume until the early twentieth century. According to anthropologists, since most Sri Lankan villages used small man-made ponds for irrigation, Bahu's dams were nothing but monuments to his potency. His less-than-subtle message was that he was so untouchable—so macho, in modern-day usage—that he could even conquer water. Unfortunately, Bahu's actions were more than just an isolated result of an oversize ego. Many leaders after him—in free nations and dictatorships alike—would use the control of water to magnify their public images.

In a classic example closer to our time, Josef Stalin after World War II built a series of hydroelectric dams in western Russia and Ukraine under a project that he touted as the Great Stalin Plan for the Transformation of Nature. It succeeded in turning the Volga, Europe's longest river, into a chain of spindly, polluted reservoirs. But Stalin had company. During the preceding decade or so, Franklin Roosevelt had ordered the construction of massive dams on some of America's greatest rivers, including the Colorado, the Columbia, and the Tennessee, changing their course forever and cementing his legacy as an economic master planner. Not to be outdone, Mao Zedong, in the Great Leap Forward of the late 1950s, put his name to the greatest burst of dam building in history to meet unreachable targets for China's gross national product.

If one were to list the most bizarrely outlandish displays of water management in history, near the top of it one would have to put the system of dams and aqueducts built in the first half of the twentieth century to funnel water hundreds of miles to the crowded megacities in southern California, a desert region that has almost no water of its own and, as such, should be little more than a cluster of small towns. But the scale of even that poorly conceived project is minuscule compared with many of the other disruptive water management schemes that have been completed even more recently. There are 40,000 dams fifty feet tall or higher on the world's rivers now. All but 5,000 of these so-called large dams have been built since 1950. China is the world's leader, with 19,000; it had only eight in 1949, at the time of its revolution. The United States is second with 5,500 large dams, although it has 96,000 smaller dams as well.

Almost no natural body of water—that is, one unaffected by some kind of management or diversion scheme—exists anywhere in the world today. A single statistic stands out to indicate how widespread water management activity has been: worldwide, reservoirs and impoundments have a combined storage capacity of upward of 10,000 cubic kilometers, about five times the volume of water in all the rivers on the planet. That means that freshwater rivers—perhaps the most essential ecosystems on Earth because they provide water to the oceans, where most evaporation, the precursor to precipitation, takes place and because they shelter so many diverse species of marine and land animals—are essentially being managed to depletion. According to a study by the Swedish Landscape Ecology Group, *80 percent* of the water in the largest rivers in the United States, Canada, Europe, and Russia is "strongly or moderately" diverted by channels, dams, and other management schemes.

Less availability of freshwater for people, due to a destabilized hydrological cycle, is just one of the consequences of dams. Widespread pollution is also a factor, as water's drainage systems that otherwise wash natural and man-made sediment out to sea are crimped. That problem, too, cuts into the amount of water available for everyday use. In the United States alone, twenty thousand river segments, lakes, and estuaries now fail to meet national pollution standards. And while industrial and agricultural waste as well as sewage may be the direct cause of pollution, waterways strung by dams and impoundments are incapable of cleaning themselves adequately. The damage to habitat is significant: more than one hundred fish species have become extinct in just the past century, and hundreds more are now endangered.

As serious as all of this is, perhaps the most devastating effect of dams is the one that's talked about the least: the large number of floods they cause. That dams are responsible for floods at all is itself an irony, considering that one of the many reasons given for aggressive water management projects is to prevent floods. Dozens of floods occur each year; in some areas of Africa and Asia, they're so commonplace that most of the rest of the world doesn't even hear about them anymore. But many well-chronicled floods that have taken place throughout history rank among the world's worst disasters. Thousands of people were killed in floods, for instance, in the Yellow River in

China, the Gambia River in western Africa, the Okavango River in southern Africa, in rivers on the Caribbean coast of Latin America and even in small towns like Johnstown, Pennsylvania, and Buffalo Creek, West Virginia, in the United States. More people die and more property is lost as a result of floods each year than from any other type of natural disaster.

In instance after instance, floods occur when dams and their abatement systems are unable to withstand the increasing pressure from swelling reservoirs and rivers seeking their natural course in rainy seasons. Usually many years after the dam or navigation canal or irrigation ditch is built, what is essentially a plumbing backup produces a sort of domino effect. When the first barrier falls, others follow, and each subsequent one tumbles more easily from the growing strength of the suddenly unloosed water. The outcome is not just tremendous human suffering and economic devastation; the world's water supply is also seriously affected. Two-thirds of available freshwater on Earth is lost each year to floods that wash the water into the seas.

Floods are so bewildering because they tend to strike without pattern, attacking water-rich and water-poor regions alike, usually with little warning. In areas where people are starving for freshwater, to have family members killed or their few possessions destroyed by a flood is the ultimate indignity, a mocking, unfair reminder of the severity of their lives, as exemplified by the water they lack. One moment there's too little water—and their children are dying of it—and the next moment, there's too much water, with the same result. It's almost impossible for these people to make peace with that fact.

In places where water is seemingly plentiful and where turning on the tap—and leaving it running for ten minutes to fill a bath—barely merits attention, living through the ravages of a flood can be equally traumatic, producing enduring nightmares and day frights in those who have experienced it. To them, it's as if the thread separating life and death as woven by nature is suddenly clearly visible, and it's so thin that it's transparent. It's unsettling to suddenly see oneself so distinctly as living at the whim of nature, rather than in control of it, and to see water—the vital element of life—transformed into an impregnable adversary.

Primarily for that reason, above all natural tragedies such as earthquakes, fires, or twisters, floods produce the worst incidents of postdisaster shock, according to psychologists and trauma experts. But until you examine the raw details, the full meaning of this phenomenon doesn't register. The level of depression and despair and, in some instances, the breakdown of order that occurs in the aftermath of a flood is startling.

Mozambique is one of the most lurid examples. Virtually every year massive floods torture this tiny nation on the Indian Ocean in southeastern Africa—a particularly harsh fate for a country whose freshwater supply is so polluted and whose infrastructure is so nonexistent or so poorly maintained that each resident on average receives only 7 liters of water per day.

The floods hit hardest in the area directly downstream from the Cabora Bassa Dam, which was built by Portugal in the late 1960s when it controlled Mozambique. With this project, the Portuguese wanted to harness for profit the powerful Zambezi River, before it was "wasted" out to sea, by producing hydroelectricity and then selling it to South Africa. In 1975, just a few years after the Cabora Bassa was completed, Portugal was ousted from Mozambique, and during Mozambique's many civil wars since then, the dam has been a flashpoint for sabotage. As weakened as the dam now is, when the Zambezi swells in rainy seasons, the Cabora Bassa is unable to hold back the overflowing river, and it lets loose a torrent of cascading water that envelops towns like Caia and Chioco.

In the floods of 2000, the scene was extraordinary. When the dam's spillway broke open, the deluges happened so quickly that even those Mozambicans who were able to get out of their homes in time to avoid drowning there sought safety in treetops. Thousands of people were perched in the branches—moaning could be heard for miles around—and they watched in terror as the waters crept higher and higher up the brown bark. In one town, residents were stranded in high branches for five days and survived by eating the carcass of a dead bull and drinking the muddy floodwater that lapped against their toes.

In all, two thousand Mozambicans were killed in floods that year and 2 million others were displaced. Mozambique lost at least a third of its staple maize crop, a quarter of its agricultural land, and 80 percent of its cattle. And when it was over, the country was dazed; residents appeared unable to piece their lives back together. The social structure broke down. Alcoholism and drug use increased significantly. The country was rife with "unusual behavior," as Foreign Minister Leonardo Simão termed it. A woman bit off the ear of a child; children separated from their parents refused to speak or eat; and people were killed and others trampled as they fought over clothes and water supplied by relief groups.

It's impossible to avoid the dismaying conclusion that in our thirst for profits, commerce, and power, our ignorance about water has been as deep as the ocean. Water is the most important, essential, and thus uncompromising natural force on the planet; there's more of it, and it's more critical to the survival of every form of life, than anything else in the world. Water, stated simply, is the difference between existence and death for every species. It doesn't matter whether the cause has been a lack of imagination and insight, or a stunning distance from nature, or plain hubris and arrogance, but our miscalculation that we could arbitrarily and continually manipulate water for our purposes without severe reverberations was—and still is, because we continue to compound this mistake—dangerously foolish.

I was probably the only person thinking these gloomy thoughts at the Hoover Dam one November afternoon, a time of the year when the desert sparkles like the reflection of light in deep black sunglasses. Most of the other people at the dam—and there were hundreds—seemed transfixed simply by being in the presence of something so much larger and more powerful than anything they had probably ever seen. Of the millions of visitors who come to the Hoover Dam each year, probably no more than a handful realize that its sole purpose is to overwhelm and corral a huge free-flowing river that is the drainage basin for an area of 246,000 square miles—or consider whether building the dam may have been an unwise decision.

And who could blame them? At first sight, the Hoover Dam is indeed remarkable, even awe-inspiring. A 726-foot-high concrete barrier straddling the Colorado River from Arizona to Nevada, at its dedication in 1936 it was the tallest dam in the world. With its perfectly proportioned art deco spillways accented by soaring arches, the geometric southwestern Indian designs in its terrazzo floors, and the two thirty-foot bronzed Winged Figures of the Republic statues at its entry point, the Hoover Dam is an artistic and engineering marvel.

President Roosevelt was effusive at the Hoover Dam's opening, saying that he came, he saw, "and I was conquered as everyone would be who sees for the first time this great feat of mankind."

By contrast, he was brutal in his description of the Colorado River and its surroundings: "Ten years ago the place where we are gathered was an unpeopled, forbidding desert. In the bottom of a gloomy canyon, whose precipitous walls rose to a height of more than a thousand feet, flowed a turbulent, dangerous river."

Taking this a step further, at another point in his speech, Roosevelt said: "As an unregulated river, the Colorado added little of value to the region this dam serves. When in flood the river was a threatening torrent; in the dry months of the years it shrank to a trickling stream."

While speaking of the Colorado's vicious floods, Roosevelt neglected to point out that much of the blame for perhaps its most startling deluge can be laid to another water management scheme that went awry. In the early 1900s, U.S. engineers, looking for ways to bring water from the Colorado to fledgling farms in California's Imperial Valley, decided to create the New River and the Salton Sea as irrigation ponds. To do this, they cut into the Colorado's banks and opened up a fresh intake, through which they hoped to send a small portion of the river westward to California growers. But the engineers made a fatal blunder. Instead of just a controlled stream of the Colorado wending its way west, the entire course of the river, drawn by gravity, raged through this new pathway. Water spilled over the artificial ditches, flooding uncontrollably over land as far away as 150 miles, before filling the New River and Salton Sea basins and eventually returning to its own natural course.

Roosevelt wasn't about to bring up this gaffe—and its role in the Colorado's flooding—at the Hoover Dam's inauguration. Instead, he spoke about how the Hoover Dam would be good for industry, for farmers, and for desert communities. But he said nothing about the damage that can occur when a natural environment is transformed into an artificial one. The dedication had one message: that the Hoover Dam had defeated a fifteen-hundred-mile river, the centerpiece of the riparian ecological system in the Southwest—something that, as Roosevelt put it, "with the exception of the few who are narrow visioned," people were proud of.

Before the dam was built, the Colorado was indeed a wild and unruly river. It regularly broke down its own banks as it created and recreated the shardlike, equally untamed landscape of the desert. But it was completely vanquished by the Hoover Dam and the subsequent free-for-all by seven states that clamored to use the river for irrigation, hydroelectricity, and drinking water. There have been few victories of people over nature as thorough as this one. Now the river is so meek that only a relative trickle reaches its mouth at the Gulf of California, near Yuma, Arizona. Lake Mead, the Hoover Dam's reservoir, can bottle up 9.2 trillion gallons of water, or nearly two years of the Colorado River's annual flow.

The environmental consequences of the Hoover Dam are as severe and far-reaching as any that have occurred anyplace in the world. For one thing, the Hoover Dam made it possible to build aqueducts to carry the Colorado's water hundreds of miles away to the desert towns of southern California. That diversion, in turn, encouraged the questionable conclusion that no real limits needed to be put on population growth in the region, and it led to the development of sprawling metropolitan areas like Los Angeles, San Diego, and Palm Springs. Without the imported water, these cities would be almost uninhabitable.

Much of the Colorado, which now has as many as fifty dams up and down its length, replenishes so weakly that the water downstream is polluted; in many places, it's mostly thick, green and murky, and its oily topping seems to be flowing sideways, not downstream. As a result, the river's once-vibrant delta in the Gulf of California has become a barren wedge of desert and salt flats, its biologically rich wetlands a thing of the past.

As a habitat, the Colorado is now virtually useless. More than one hundred plant and animal species that live in the river are endangered.

I spent five hours at the Hoover Dam and spoke to a dozen or so visitors as well as some Department of Interior employees about my suspicion that the dam might be a monument to our folly or, more dangerously, a testament to our misguided notion of environmental invulnerability. Nobody wanted to talk about this issue, not even to argue the other side and explain why I was wrong. Parents held their children closer to them, seemingly trying to protect them from my questions. The kids weren't listening anyway. And Hoover Dam staffers told me to speak to the press office in Washington, D.C.

The only person willing to even come close to addressing the topic was a tour guide, a woman in her fifties who said she grew up in the countryside south of Boulder City, Nevada, where the dam is located. Her earliest recollections are of learning how to swim on the Colorado River, which she described as moving so fast then that at times, while riding in the belly of a truck tire while her grandfather fished nearby, she felt as if it were going to carry her away.

"The Hoover Dam is like Disneyland," she said. "It's an amusement park, a place to come to when you want to see firsthand and believe, whether it's the truth or not, in the power of man, not in his faults. Nobody at Disneyland talks about the puny wages that the person making the T-shirts somewhere in who-knows-where is getting. And here we don't talk about the Colorado River."

She was right. Nobody does talk about the Colorado River here, unless it's to emphasize a point about the greatness of the dam. The Colorado has become a straight man, the Hoover Dam's captive sidekick. I felt a strong pang of sadness, standing on a perch at the top of the dam, listening to the weak lapping of Lake Mead, the embodiment of the Colorado River's defeat, as it tapped lightly—like a faint, near-dead knock on a door—against the walls of the inviolate structure.

The ancient Egyptians were one of the few civilizations that structured their day-to-day, month-to-month activities of life to match the patterns of a powerful and extremely complex water system. The Egyptians accomplished this significant feat by looking at water—and thus water management—through the prism of nature and nature's relationship to their gods.

It was an extraordinarily sophisticated environmental worldview, no less impressive six thousand years ago than it would be now, because it enabled the Egyptians to reject the belief, common even then, that controlling water was the only way to satisfy a society's commercial motives.

Most remarkably, the Egyptians understood the value of floods and viewed them as more important than dams. Instinctively, they grasped a guiding ecological principle: that floods are an essential part of water's activities on Earth, organic events driven by the necessity of rivers to seasonally spread out and then retreat back into their basins. The essence of a river is that it's dynamic, always in flux: eroding its bed, depositing silt, seeking a new course, bursting its banks, drying up.

Each year the Egyptians had a front-row seat to view one of the world's most convincing demonstrations of this fact. Cutting through the heart of Egypt is the Nile River, which at nearly 4,200 miles is the longest river in the world. In ancient times, for much of the year, the Nile was a relatively calm waterway as it flowed northward past Cairo into a turkey-necked delta and then out to the Mediterranean Sea. In these seasons, most of the river's water came from the White Nile, so called because of the pale color of the muddy liquid that it carried from the big African lakes as far south as Kenya and Uganda.

But in the deepest summer, when Egypt is as stark a desert as any the world knows, the Nile would be overrun by the beautifully crystalline waters of the Blue Nile, which poured out of the Abyssinian Mountains in Ethiopia as the snows melted. Its waters would transform the Nile into a fierce rushing river that stormed unimpeded into the hot, dusty landscape of Egypt, overflowing its banks and drowning the narrow farmlands running along its length on either side. As it flooded, the river would pick up bits of soil and plant life. Then slowly it would recede, leaving behind layers of nutrient-rich silt and food in the form of dead catfish, mullet, bolti, and perch. As a result, the deep black soil along the Nile was exceptionally fertile. Lying in a delta that totaled only about 3 percent of Egypt's landmass, it supported almost enough crops to feed the entire country.

To the Egyptians, who interpreted everything in terms of the relationship between natural and supernatural forces, this yearly cycle of the

Nile—and the incredible sustenance it brought—was nothing short of a miracle: a gift of the goddess Isis, whom they associated with rebirth. In the metaphors of Egyptian mythology, the Nile floodwaters were the tears Isis shed for her husband, Osiris, whom she resurrected after his murder by a jealous god. With such divine origins, the Nile, even at its most devastating, was obviously not to be tampered with. Building dams and other abatements was out of the question. The Nile was allowed to freely spin through its cycle, engorge with water from high in the hills, replenish the earth with its revitalizing liquid, and then return to its tomb in the mountains. To keep out of the river's way, people farmed on its banks but lived some distance beyond the point where its floodwaters crested.

This simple but extremely successful approach to water management ended in 1970. That year Egypt's founding father and longtime president Gamal Abdel Nasser, in one of his last official acts, ordered the building of the Aswan High Dam, a series of huge locks and gates on the Nile that trap the river water in a reservoir six hundred miles south of Cairo. By taming the Nile and setting up a network of irrigation ditches, Nasser reasoned, water could be distributed to farmlands throughout the year, even during droughts. The growing season would be extended, ensuring enough food for Egypt's rapidly multiplying population.

But the Aswan High project had serious flaws. Interfering with a water cycle as orderly and natural as the Nile's should never be done lightly, and Nasser ordered no independent environmental impact studies to determine the prospective repercussions of the project. One reason is that Nasser had another motivation for wanting to construct Aswan High: to ensure that he'd be remembered in Egyptian history long after he died.

Much like the great temples of the pharaohs (many of which were submerged by the building of Aswan High, while others were moved at great expense), this project was built to be noticed. The dam itself is nearly four hundred feet high and more than ten football-field lengths across. The reservoir impounded by Aswan High, called Lake Nasser, is one of the largest in the world, covering an area three hundred miles long and ten miles wide. The cost of the project finally totaled more than $1 billion, a high price for a nation in poverty.

Nasser stopped at nothing to make sure the project was completed. During a low point, when the United States refused a request for additional funding because Nasser had gotten too friendly with the Soviet Union, the Egyptian president offered the Soviet Union almost all the revenue from the Suez Canal and a permanent military presence in exchange for $300 million to finance construction.

Aswan High had its formal opening in 1970, just a few months before Nasser's death. And with that, the Nile's natural pattern of summer flooding—a yearly cycle dating back, no doubt, to prehistoric times—ended forever. As Nasser had promised, Egyptian farm output increased significantly. But the attempt to control the river produced serious side effects. Silt levels in Lake Nasser rose dangerously, because the Nile was trapped, setting the stage for widespread pollution in the reservoir and subsequently on the farmlands irrigated by this water. Meanwhile, to replace the nutrients that used to be carried by the Nile, downstream farmers overfertilized their land. When it rains, much of this fertilizer washes into the river and ultimately flows to the Nile's delta. With much less water available in the river to filter out the impurities, the entire lower Nile basin is also becoming polluted.

Moreover, without the sediment from the overflowing Nile to compact the land adjacent to the river, downstream banks are eroding. If this process isn't contained—and there are no plans or money to do so right now—then there will likely be floods on the Nile far more dangerous and powerful than those that used to occur seasonally. As a result, the gains in cropland upstream could eventually be diminished by the acreage destroyed by deluges downstream. And with the makeover of the Nile's natural course, the fish population in the river—even commercial catches like sardines and shrimp—is dropping rapidly. So Aswan High, intended in large part to increase the amount of food available to Egyptians, may actually reduce it over the long term.

On a more profound level, with the completion of Aswan High, Egypt cast off a unique relationship with water that will never be recaptured. It severed the bond between the Nile and the people who relied on the river for their lives and their livelihood and bartered it, it could be said, for

Nasser's $1 billion legacy. What had made the connection between the people and the river so singular was that the Egyptians were not afraid of the Nile's floods; to the contrary, they welcomed them. They saw floods not as nature out of control—let alone as nature in need of control—but as a gift to be treasured.

This worldview is extremely unusual. Very few civilizations have understood that respecting water's natural course is more valuable and safer than rejecting it. The failure to understand this principle has had a devastating cumulative environmental effect and has enabled us to blindly believe we can remake our environment as we wish.

By pure coincidence, I arrived in Rapid City in June 1972, just a week after the Canyon Lake Dam collapsed and what's now known as the Big Flood occurred. At the time, I was a musician performing with a rock-and-roll band, and we were in the middle of a cross-country tour, driving a GMC van from city to city in the northern United States. We were supposed to play a show in Rapid City, stay a few days with some local people, and then go west to California. We ended up spending a couple of weeks in South Dakota, caught up in the residents' utterly fascinating attempt to search for explanations for the flood. None were forthcoming.

Upon first entering Rapid City, I was struck by its solitude. It had the look of the worst type of ghost town—the kind where the people are alive but the streets are dead. A hushed breeze skimmed over the vacant roads, carrying strafed pinecones torn out of the Black Hills. I kept hearing soft gusts that sounded like cars approaching from a distance, but when I looked around, the streets were virtually empty.

Here and there the inanimate signs of the disaster's aftermath jarred unexpectedly. A doll's head, bald and severed from its body, lay in the mud near the now torpid creek that cut through the town center; one gray eye was opened in a lifeless stare, the other shut tight.

The thickness of the air was suffocating. The city was so drenched in humidity it seemed to be perspiring, as water in the streets evaporated in the bright sun. There were few people outside, but in the distance came the sounds of cleaning up: the deep motorized rasp of power tools and then the thud and clank of wood and metal being tossed onto piles.

Downtown, rows of stores had been boarded up with hastily cut sheets of plywood. In many cases, these attempts to save the storefronts had come too late. Beneath the thin plywood, which was mostly waterlogged and transparent, broken windows and shattered displays were visible.

Rapid City, on the far edge of the Great Plains, is a small town of mixed identity, neither entirely Midwest nor Wild West. It was settled as a mining center in 1876, after gold was discovered in the nearby Black Hills. That history gave Rapid City its initial reputation as a lawless western village of prospectors and, soon after, of loggers and cowboys.

But topographically, Rapid City isn't Big Sky country. With its softly paced hills and wide-open agricultural fields, it looks more like the breadbasket communities in Missouri and Minneapolis. This disconnectedness, poised uncertainly between East and West, difficult to categorize and place, has made Rapid City a mostly forgotten American town. It's known more for being close to Mount Rushmore, the Badlands, and Little Big Horn, or to the farms of Minnesota, Nebraska, and Iowa, than for being the center of anything.

That's one of the perplexing aspects of the 1972 flood. No one would have wagered that of all midwestern towns, Rapid City would be the target of a flood of that magnitude. For one thing, Rapid City is anything but a river town. The only water that runs through it is a tiny trout stream that ambles along at a pace of a few hundred cubic feet per second. Rapid City is situated a few hundred miles west of the Missouri River and even farther from the Mississippi, both of which overflow their banks many times a year. It seems unfathomable that the town's one small stream would suddenly be transformed into a barnstorming, swollen river that took oxbows at speeds five hundred times its usual pace—that it would trump the big rivers of the Midwest in the severity of its flood. But midwesterners still cringe when they consider the enormity of the flood's toll. Two hundred thirty-six people were killed and three thousand others were injured in a town of only 75,000 or so. One out of every ten homes was destroyed, and tens of thousands of head of livestock, horses, dogs, cats, and other animals were carried away, buried at the flood's mouth.

As I drove through town in the aftermath of the flood, the general store was one of the few shops that was still open. The proprietor, a gaunt, sad-faced, wan man named Chuck, with Peter Lorre eyes, was prone to gallows humor. I asked him how he had been able to clean up and reopen so soon after the flood, when everybody else was still struggling. "Disasters are good for business," he told me matter-of-factly.

Chuck's store usually doubled as a local social club. During good times, barrels lined with burlap stood near the counter, holding staples like feed, sugar, and flour, and doubling as leaning posts. People would linger there to engage in idle chatter, to avoid going back to work or to an empty house.

The day I visited the general store, though, no conversations were taking place, although there were about twenty people in the store, browsing the shelves and aisles. In fact, the shoppers, some of whom must have known each other, weren't even making eye contact or greeting each other. Virtually the only sound was the shuffle of boots on the hardwood floors, which were covered with sawdust to soak up the water that darkened the slats in uneven patches.

Chuck, however, was loquacious, and this librarylike stillness was getting on his nerves. He talked on about a variety of things, including his view that one good thing that came out of the flood was that the people of Rapid City found religion. There were more cars in the parking lot of the churches this past Sunday, he said, than on the speedways and at the drive-in movies the night before. That's the good part, he said, but something's still missing: "We got spiritual, but now we have no spirit.

"You want to really know why it's so deadly around here now," he continued, leaning forward on the counter, and speaking to me in a hush, as if he didn't want to talk about his neighbors in front of them. "It was a broken promise. We never had a flood around here. Never. Rapid Creek always seemed so lazy. Nobody ever expected her to destroy her own banks.

Let's face it, the creek itself was barely on anybody's mind—we hardly noticed her anymore.

"Then last week she changed the plan, suddenly, without warning. It was like we had an arrangement one day, then the next day we didn't. People are still trying to comprehend how that can happen. How the rules can change so quickly. And we're trying to get used to the fact that a couple of hundred of us were wiped out in no more than a few hours. Vanished like they were chosen to die. But there's no pattern there either. The good are gone and so are the bad. It's just the damnable mystifyingness of it all."

Chuck took off his sweat-stained glasses and slowly pulled a crumpled handkerchief out of his pocket. He wiped the lenses clean. When he had finished, he tucked the hooked ends of his silver wired frames back behind his ears, then looked at me with a dazed expression, like someone who was returning from having lost possession of himself for a moment. He couldn't remember why I was in his store. "What did you come in here for, again?"

"I need directions to Dark Canyon Road."

Thirty years after the Rapid City flood, we still have no easy explanation for the eight-hour rainstorm that produced it, a storm so unprecedented in the region that it seemed otherworldly. In fact, that this meteorological phenomenon has not been repeated only magnifies the mystery.

In the days, weeks, and even years that followed the disaster, numerous suggestions were offered to explain the storm and the flood. Perhaps an unfortunately timed cloud-seeding by the South Dakota School of Mines, part of a precipitation study, had something to do with the torrential rains. Or perhaps the massive dams on the Missouri River—the most straightened and locked waterway in the world, with billions of dollars of braces marking almost every mile of its journey northeastward from Montana to St. Louis—played a part in the sudden swelling of Rapid Creek, one of its tributaries. Or perhaps it was the work of an Indian chief, his name long forgotten, who a century before had prayed loudly to the skies for pain and destruction to rain on the white man, soon after settlers broke agreements and forcefully snatched some of the Sioux's most sacred territory in Pine Ridge, Deadwood, and Wounded Knee.

Recently, teams of hydrologists and meteorologists throughout the Black Hills have been trying to recreate the events of June 9, 1972, on super-computers. Most of their work has centered on building three-dimensional models that simulate the organization of thunderstorms. These digital depictions, based partly on weather information gathered in Rapid City the night of the flood, have reconstructed the rapid updrafts that led to cloud and precipitation formation.

But the best these scientists have been able to do so far is to piece together the singular climactic conditions that existed at the time. That they will figure out why the storm occurred remains only a distant hope.

The first event leading to the storm appears to have been the unusually high level of humidity that developed on the morning of June 9. Hundreds of miles to the east, hot, wet winds built up, then rushed westward over the Great Plains into Rapid City. By eight A.M., surface dew point temperatures, which measure moisture in the air, were in the mid- to upper sixties in western South Dakota, nearly twenty degrees above normal.

As the condensation-heavy winds cut a wide and noisy path through Rapid City, they hit an obstacle they couldn't breach: the east side of the Black Hills, the mountains that lie to Rapid City's west. With nowhere to go, the thick air was propelled upward in a vortexlike pattern. And when it hit the cold atmosphere above the Black Hills, clouds formed that quickly and progressively thickened. Eventually, they exploded into thunderstorms.

There's nothing especially strange about that scenario. The humidity was well above normal on June 9, but besides that, the meteorological activity was typical of the storm-producing weather pattern for the Rapid City area. Usually, though, the region's storms are pushed westward or northward away from the Black Hills, by a high- or low-pressure system that arrives from the south or the north. This time something strange happened. Two different pressure systems converged above the Black Hills simultaneously: a powerful one from southern Canada and the Great Lakes, driven there by an upper-level storm over eastern Canada, and a weaker one from Colorado and the Southwest. Instead of propelling the storm away from the region, the combination of these two pressure systems formed a deep trough on top of the mountains. The heavy, humid air climbing the Black Hills was trapped in this sinkhole, where it repeatedly and hungrily fed on the cold atmosphere above the mountain to produce one thunderstorm after another in an unrelenting display of climatic power. It was as if the hydrological cycle of rainfall, evaporation, condensation, and then precipitation again were stuttering, stuck in place above the Black Hills and performing over and over in minutes what normally took hours or even days to complete.

None of the tiny waterways around Rapid City could handle the deluge. Rapid Creek started rising within a couple of hours of the first storm, which began at about four in the afternoon; it ruptured its banks by eight o'clock. The first damage was mostly to cars, which stalled out, their passengers trapped in the water that filled the streets. Then the homes nearest to the creek were hit as water poured into basements and then climbed through floorboards and windows into ground-level rooms. A few hours later, Canyon Lake Dam crumbled, and the spreading flood couldn't be stopped. The entire area was swept up in an uncontrollable cascade. Nothing—people, animals, foundations, or even concrete buildings—was strong enough to withstand the flood's intensity and might.

Researchers today view the storm as an anomaly and leave it at that. It is true that the School of Mines seeded the clouds the day of the flood, but the scientists discount that as playing a role. The amount of salt deposited in the air wasn't nearly enough to produce the storm that occurred. Besides, the clouds had been seeded many times before without similar consequences. As for the dams on the Missouri River or Sioux entreaties to their gods—the echoes of the Ghost Dance ritual that was practiced and then outlawed in the late 1800s, whose purpose was to rid the region of white settlers—researchers decline to connect these with the flood either.

In the three decades since the flood, Rapid Creek has been reinforced with a wide floodplain, strung by a bike path that runs through the center of town and into the hills. No new homes have been built close to the creek. The huge gates that splintered when Canyon Lake Dam failed were never reconstructed: now, there's a free-flowing spillway. The city is braced for another cataclysmic downpour, but it probably won't happen again for hundreds of years.

Economically, Rapid City has grown much stronger since the flood. Dozens of new shops and restaurants have replaced the waterlogged stores. In the process, the city lost much of its tired Old West look. With its bucolic surroundings and its low-pressure lifestyle, Rapid City has attracted thousands of new high-tech, service-industry, and agriculture jobs. Unemployment rates have plunged to below 3 percent.

But a question persists about the flood: How much fault lies with the massive water management projects that had transformed the Missouri River into a controlled waterway? Did the Missouri's complex man-made plumbing system back up Rapid Creek, the largest tributary of the Cheyenne River, which in turn discharges into the Missouri, near Pierre, South Dakota? Researchers may discount this possibility, but Rapid City residents and many environmental scientists are still suspicious that the Missouri played a part.

The Missouri, which was brought to public awareness when Thomas Jefferson commissioned Lewis and Clark to explore it in 1804, is a fascinating river. For one thing, it's one of the world's longest, spanning 2,400 miles between its headwaters at the confluence of the Jefferson, Madison, and Gallatin rivers in Montana and its mouth just north of St. Louis, where it joins the Mississippi River. It's almost impossible to drive through the Midwest without repeatedly crossing the Missouri.

And only the most jaded would fail to be impressed by the expanse of the river's long-armed, octopuslike basin. It stretches out over nearly a hundred tributaries, encompasses more than 300 million acres, and drains about one-sixth the surface area of the continental United States.

The Missouri as it exists today is divided into approximately three equal sections. The lower third, the part that lies below Sioux City, Iowa, is straight and narrow—"channelized," in the language of flood control experts. The middle third, north of Sioux City, is impounded by six large dams. And the upper third, in the river's highest reaches, consists mostly of regulated "free flowing" stretches of water. Only one percent of the river's length is uncontrolled by humans.

The suspicion that the damming of the Missouri was responsible for the flooding of Rapid Creek won't die because four of the river's six dams are located in South Dakota. The largest is a rolled-earth megalith that forms Lake Oahe near Pierre, just fifty miles from Rapid City. No matter how powerful and unusual the storm was, the hunch still persists that if the Missouri had not been so relentlessly contained by dams, the swollen Rapid Creek could have emptied into the river's tributaries without having to overrun its banks. That hunch isn't unusual: when any environmental disaster occurs in the Midwest, the Missouri has become a usual suspect.

All of the attempts to redirect the natural flow of the Missouri have produced a river that, even many ardent anti-environmentalists admit, is most fittingly described by Chad Smith, a coordinator for the American Rivers Group: "It's a mess. There's no plan. There's never been a plan. People just got their hands on it and ruined it."

Floods were once an essential part of the Missouri's ecosystem, critical to the river's management of its own environment. By overflowing its banks seasonally—usually in spring, when rainfall was high and the winter's snow and ice were melting off of the mountains—the Missouri replenished its backwaters and maintained its braided channels, islands, and sandbars. For thousands of species, the rejuvenated sheltering crevasses of the river's sloughs, oxbows, mudflats, and marshes triggered their reproductive cycles. What's more, the floods provided an exchange of nutrients between the floodplain and the river.

Now the Missouri appears to be at its most natural when it's actually at its most artificial. Stand at sunset above Lake Oahe, the reservoir formed by Oahe Dam, and you can watch a shower of bright yellow stripes in the sky reflect and illuminate the vast darkness of the meekly rocking lake below. Listen to the stillness, the complete quiet except for the seesaw sound of the water, the echo of the terns' call, and the jump of the walleye in the lap of the river. It's torturously serene. And it's easy to forget, or to not care, that it's man-made, lacking all of the Missouri's original rough-hewn personality.

Ironically, while the Missouri was remade in large part to prevent it from flooding, the river still floods often. The dams and channels make the

Missouri deeper and faster-flowing than it was originally, and during a heavy rainy season, it crests much higher than it used to, easily overspilling levees that were built to protect the farmland and cities from flooding. Some of the worst floods have washed away hundreds of homes and belongings. In a few cases, whole towns have had to move away from the river.

But the attempt to control the Missouri has had its biggest impact on those species that once relied on the river's natural conditions. Thirty-three of the 156 fish species native to the Missouri River basin have now been placed on state and federal endangered species lists. Another seventy have been tagged as rare. Big river fish such as sturgeon and paddlefish, which used to be common in the Missouri, have been greatly reduced in numbers.

Only one family on Dark Canyon Road, the Lockharts, survived the Rapid City flood. Following the directions I got from Chuck at the general store, I drove up the sharp incline into the mountains, out of the town's business district toward the countryside. The Lockharts' home stood on two hundred acres at the very top of a bluff in the Box Elder Mountains. Every house below the Lockharts' that I passed had been lifted and moved by the flood a week earlier, coming to rest askew and ripped apart; none were unscathed, and no people were visible.

Land rich, but cash poor, Jim Lockhart, the father of ten children, earned a meager salary driving a freight train. The property was an inheritance. His great-grandfather had been deeded the highland acreage more than a hundred years ago, when he was one of the first sheriffs of Rapid City.

Jim was an outsized, boisterous man, full of virile energy, who had proved to be something of a hero during the flood. As the water headed toward the Lockhart home—it crested no more than a hundred feet below the property line—Jim and his seventeen-year-old son, Jim Jr., worked their way down the hill to try to save their neighbors. In hip boots and plastic coveralls, they literally grabbed people as they were being thrown into the maelstrom, desperately trying to swim when swimming was virtually impossible. During an exhausting hour or so, Jim and Jim Jr. had been able to save a handful of people; the rest had been ripped away and sent tumbling in the cascading water down the mountainside.

Jim's wife, Teena, said that her husband knew he had done all he could to save his neighbors, so he wasn't tormented about what happened. He went back to work as soon as the trains were running again. But she couldn't shake it. A Rosebud Sioux, Teena grew up in these hills and learned there what it meant to seek balance with one's environment. She was taught to respect the forces of nature, because they were powerful; only those who showed deference would get respect in return.

She was soft-spoken and slight, with long flinty gray hair worn in a bun. For the first few days that I stayed with the Lockharts, Teena spoke little about the flood; it was too fresh in her memory. And her difficult-to-bury feelings that the flood might have been nature's angry response to the way people have treated the waters in Rapid City and nearby were too uncomfortable to think about, much less put into words. Her other contradictory thoughts were even less consoling: the flood might be a sign that nature lacks order or logic, placing in question the long-held beliefs that she and her ancestors lived by. Teena finally spoke to me about this during a walk we took through the woods toward a high ridge. It was eerily quiet—we heard just the sounds of birds and our footsteps trundling the wet soil. When we reached the clearing, on the ridge, an extraordinary sight took me aback. In the hills surrounding us, the exact spot where the flood had crested was dramatically recorded by remarkably distinct tree lines and water marks. It looked unnatural. The usually dark, overgrown Box Elder Mountains were treeless except for the very top, where there were thick stands, looking like lone wind baffles next to an ocean. Right below these isolated clusters of trees, the mountains were as brown and barren as the Badlands. Unmistakable sootlike water marks circled the Box Elders, measuring the relentless climb of the flood until it reached its high point.

Teena said the Box Elders looked strip-mined and alien to her now. The mere sound of the echoing wind crashing against the bare mountains frightened her.

"I used to sit on my porch with the kids at night, reading to them or rocking them and feeling enveloped by the mountains," she said. "Now they haunt me. Those water marks recall the rings of the dead that we put on trees in the reservation, when relatives passed on."

One of Teena's most difficult memories was the reaction of her youngest children—a five-year-old and a seven-year-old—to the flood. "They were dancing, like kids do, in the rain, which was coming down in torrents," she told me. "And not understanding what was going on, they were laughing at the people and animals swept up by the flood. It looked like a giant water slide to them. It looked to them like the people that were dying were having fun."

Teena said she had to decide whether to scold the children for their out-of-place behavior or ignore it, since they didn't understand the seriousness of what was going on.

"I chose to leave them their innocence. You only have that once," she said. "Just because I was losing mine didn't mean that they should lose theirs."

Maybe, I said, the flood is the answer to the Indian Ghost Dances, the retribution, wrought finally after all these years, against the imperious white settlers. If so, then the flood and its aftermath would fit the notion that there is a logic to nature's actions, as she was brought up to believe.

"Even you—and you've just been here a few days—are searching for answers. You're looking for why so badly that you're willing to consider some old far-fetched Indian legend to explain why so many people are dead," she said. "Well, I guess if white people are willing to believe it, I'm not going to dissuade them. It's not the worst thing for more people to fear our witchcraft."

Teena's point, of course, was that even if there is a logic to nature, there are still no rational explanations that people can comprehend for natural disasters like floods, so why seek answers? The fact is, the only thing that we can really learn from floods is that they—and more important, water—are mostly out of our control. No matter what we do, no matter what kind of water management projects we attempt, we won't be able to eliminate floods, and often we'll worsen the situation. Still, this is such a difficult reality for people to accept that we ardently and often preposterously resist it.

In spring 2001, circumstances in Davenport, Iowa, drove this point home. Flood waters were cresting up and down the Mississippi River, and damage to houses and businesses was mounting. Unlike most other towns and cities on the river, Davenport didn't have a flood wall. Instead, during the prior decade, it had developed a parklike waterfront that depends on the river for much of its attraction, and had moved virtually all homes and businesses as far from the water as possible. Davenport officials justified this decision by saying that flooding is an inevitable part of life when you live on the Mississippi, so they chose to build their city around the possibility of floods and not try in vain to stop them.

Davenport has had plenty of critics, though, including Joseph Allbaugh, the head of the Federal Emergency Management Administration, who complained loudly during the 2001 flooding that "all of this could easily be prevented by building levees and dikes." That is one of the stranger statements about water that I've come across in my research. Allbaugh made those comments when almost every Mississippi River community within driving distance of Davenport had been overrun by the floodwaters. And practically all of these places had invested hundreds of millions of dollars in flood control structures so big that the communities were effectively walled off from the river that was their birthright—and still they were suffering significant flooding and significant damage. Even supporters of the water management projects were embarrassed, or amused, by the FEMA chief's scolding of Davenport.

To hear Davenport being chastised for its approach to flood control when no city or town had been able to avoid the Mississippi's waters, no matter what measures had been taken—and some sustained even greater losses than Davenport—seemed ludicrous. Yet perhaps it wasn't as absurd as it sounded. Allbaugh's attitude was typical of one we've always held, in foolishly thinking we can restrain raging waters. It occurred to me with some disappointment that when people are determined to avoid the obvious, not even a flood—or thousands of them—can get their attention.

2

Invisible Desert

There is a silence where hath been no sound,
There is a silence where no sound may be,
In the cold grave—under the deep deep sea,
Or in wide desert where no life is found.

THOMAS HOOD, "SONNET: SILENCE"

DESERT CENTER CLINGS to the exit that bears its name on Interstate 10 in southeastern California with such resolve, it's as if it fears being drawn into the uninterrupted wilderness behind it—and ending up even more invisible than it is now. The town's entire population, about 130 people, lives in fewer than fifty homes clustered close to the highway—tiny ramshackles, painted every pastel to repel the burning sun that moves in slow motion over the wide-open, almost lifeless terrain.

Seemingly out of character in this low-tech zone, there's a satellite dish in almost every front yard, even though most of these lots are minuscule, about the size of a pickup truck. Broadcast signals and cable don't reach as far as Desert Center, so a satellite dish is a necessity. Without TV, people would be hard-pressed to find anything to do here. Desert Center is a town without a core. The nearest movie theater, bowling alley, shopping mall, and library are at least fifty miles away.

But it does have I-10, and in many ways Desert Center best expresses the personae of that highway's towns. Beginning at Los Angeles, I-10 lumbers eastward across southern California, intersecting thriving cities like

Palm Springs and Indio and forgotten ones like Blythe and Plomosa, cutting a straight path to the Colorado River at the Arizona border. All of the places on these I-10 exits—from Los Angeles and its sprawling network of attached communities to the empty California badlands to the east—were originally, as Desert Center still is, bleak desert towns. No matter how they look now, they once were mirror images of Desert Center, fashioned out of bone-dry dirt that in its natural state was able to sustain little more than tortoise-paced cactus growth.

That heritage is no longer obvious in metropolitan Los Angeles. The green lawns, the groves of trees in the hills overlooking the Pacific Ocean, the backyard pools, the decorative fountains in village squares, the children's wading holes, the persistently hissing water sprinklers in even the most urban of settings—all camouflage and provide relief from the surrounding oppressive desert environment. Looking at them, it's easy to forget about the age-old barren, water-scarce geology that overshadows the current decades-old Los Angeles epoch.

In fact, not until you get about ninety miles east of Los Angeles, near Palm Springs, does the desert landscape truly assert itself. At that point, without warning, the influence of Los Angeles wanes and the backdrop to I-10 becomes nothing but miles and miles of brackish, raw-faced mountains, which from a distance appear to be as weightless as sand hills, separated by burnt brown valleys. A few stripped shrubs, little more than skinny twigs linked in patterns that look like nerve endings, are barely visible. The view from a car is so monotonous and flawlessly repetitive, it's as if the vehicle is standing still and the background is flashing by in a loop, the way Hollywood used to film driving scenes in indoor studios.

Every once in a while the predictable rhythm of this panorama is broken by sparkling twenty-foot-high, fat, steel-gray aluminum pipes that emerge suddenly from near the tops of the mountains, slide down the slopes, and then burrow into the ground. There are usually three pipes within a foot of each other, each exactly the same size with the same elbow bend. These are aqueducts that transport water, mostly from the Colorado River but some of it from northern California, hundreds of miles to Los Angeles's Metropolitan Water District, which disperses it to 17 million people in twenty-seven cities in Los Angeles, Orange, San Diego, Riverside, San Bernardino, and Ventura Counties. Without access to this imported water, southern California—with its famed cities like Los Angeles and Palm Springs—would not exist as we know it today. There's not enough local water in southern California to support anything more than a dusty prairie town, the type you'd see in *High Noon* or *Unforgiven*. The pipes surfacing from the mountains are a cinematographer's trick, like strings that suspend a flying actor; they're the hidden reality that allows southwestern California to maintain the illusion of being something besides a waterless desert.

By the time the Desert Center off-ramp comes in sight—about three hours east of Los Angeles—the population and the number of towns have so thinned out that the exits are few, perhaps one every twenty-five miles or so. For most people, the only reason to exit at Desert Center is to refuel, use the bathroom, or grab a cup of coffee and perhaps limp, week-old Twinkies. A combination gas station, food market, and diner called Desert Lily stands adjacent to the northern edge of the highway, open twenty-four hours, according to a torn cloth banner hanging limply between the awning over the pumps and the rectangular little building where food is served and sold.

I pulled into Desert Lily one noon in late August. A thermometer—a promotional gift from Bardol engine additive—was suspended in the shade on the pole next to the gas pump. Its black hand pointed to 110 degrees. The rumble of cars and the bouncing of truck beds careening down the highway clamored in the background. But when I turned my back to the interstate and faced north into the empty desert and the sun-drenched air, which had a brilliant white hue, the bleached-out color of a painter's mask, all noise abruptly faded away. I heard only a stillness, broken by twirling weathervanes, spun by a soft wind reflecting off the parched mountains in the distance.

Glenn Boyer, a stocky man with hunched shoulders, dressed in oily gray mechanic's coveralls, was filling my gas tank. "I don't know why you want to go to Rice. There's nothing there," he told me. "It's fifty miles straight into the desert and straight to nothing. There was a grocery in Rice, but that burned down a year ago. And the gas station burned down two months ago. Now it's a vacant town."

Yes, he said, he had heard about the water project near Rice, at the old Cadiz Lake, which was dry, of course. It was one of the more bizarre ideas for supplying water to Los Angeles and its surrounding towns, to meet the region's still mounting needs. Boyer laughed. "Do you see any water around here?" he said. "This is the desert. I can't even imagine what they plan to do, or why they think it will work. But if you're going there, do me a favor— take some water with you. It's hot, and there's no one to help you if something goes wrong."

"I have water," I said, pointing inside the car.

He eyed the liter bottle of Poland Spring in the cup holder and looked at me more with impatience than amusement. "No, not to drink. For your car. If you overheat, how do you expect to get back? Are you going to walk the fifty miles?"

I was stopped short—and disappointed in myself, as I had been numerous times during the research for this book—by how easy it is to underestimate the importance of water; it's unfortunately second nature for people to do so. Water had become close to an obsession to me, enough so that I was driving out into the deepest desert to witness for myself yet another foolhardy water management scheme, and still I was once again faced with the realization that we understand so little about it. At odd moments, though, like this exchange with Boyer, it did penetrate: there's virtually nothing that moves or lives that doesn't rely on water to subsist. And when Boyer handed me two antifreeze containers full of water, I was struck by something I hadn't completely understood before, even though in hindsight it seems so obvious: water's vital role isn't limited to living things alone; it extends to the machines we make. They couldn't exist without water either.

"I'll call if I get into trouble," I said.

"You won't on your cell phone. There are no cells out there. And there are no pay phones that work. Why should there be? Nobody lives there."

Not surprisingly, Boyer was right about the town of Rice—it was empty. Charred remains of the burned-out gas station and grocery store lay directly across the street from each other and formed two piles of rubble. Ash by ash, a quiet, tiptoeing wind was carrying away what was left like a bird building a nest. A two-lane road, with barely visible white lines down the middle and on its shoulders, cut through the center of Rice. Fresh tire streaks in the dust, made by my car, were the only visible signs of traffic.

Cadiz Lake was a few miles north on this road. It, too, was empty—nothing but a large dry hole in the ground, partially filled by rock and sediment, with rugged, uneven banks. The sole indication that something out of the ordinary was happening here—that this wasn't merely one of the dozens of dry lakes in the desert, a hole in the ground with no tangible value, but the site where L.A. was looking for its future water supply—was a towering construction sign that read "Cadiz Groundwater Storage and Dry Year Supply Program." That sign was the only suggestion that anyone had recently visited this otherwise desolate valley.

I expected to see heavy earth-moving equipment, perhaps a crawler dozer or an excavator. But there wasn't a single machine. Nor was there any clue that work had begun on the lake. Like the rest of the desert that I had traveled through, this valley appeared to be untouched territory. Surrounding Cadiz Lake was a pocked brown moonlike landscape of mountains and valleys, whose alien aspect was made even more vivid by the knowledge that besides me, there was almost certainly no one—not a single human being—within, say, fifty miles. In truth, there seemed to be no life of any sort, not just human, in view. Nothing moved on the ground; nothing flew overhead; no calls of crows or mordant homing sounds came from the air; no cactus grew in the dust. No bugs ground slowly in the brown clay or circled to settle on my skin. The only motion I saw was the feathery cartwheels of tumbleweed, one after another, making a perfect beeline for my leg, as if they could see me. These reflections on solitude—feelings that were overwhelming and inescapable as I stood isolated in the desert, staring at an empty lake bed with its incongruous construction sign but no construction activity—frightened me, perhaps because for the first time I experienced what it's like to be crushingly alone.

The absence of noise was a powerful presence. Without any outside stimulation to distract me, the disconnected, jittery riffing of my thoughts, my heartbeat, the sound of my breathing became so much more distinct. Their rhythms were hypnotizing.

Then, my attention was suddenly drawn outward. There was, I realized, a sound I had initially failed to hear, one that had the basso pitch of buzzing bees. Its almost imperceptible whir seemed to start well off in the distance and then move closer, like a cool gust of wind approaching—but none came; the desert air didn't budge at all.

At first I thought it was swarming flies, but there were no bugs in sight. The sound moved. One moment it was in front of me, then behind. I spun around, following the aural swath it was cutting. I still saw nothing and couldn't help but laugh a bit at how foolish I must look—a tiny figure in the center of a miles-long deep, hot tobacco-colored valley, twirling to catch up with a noise that seemed to always be one revolution ahead of me.

I never did find out what the sound's origin was. I later asked half a dozen people who lived in the desert—including Glenn Boyer at the Desert Lily diner—if they had ever heard anything like it, and all of them said they had. Almost everybody who goes into the empty reaches of the desert claims to hear it at one time or another, they told me, but nobody has been able to pinpoint what it is. One intriguing myth holds that it's the sound of the earth spinning on its axis; like stars that are invisible in well-lit urban areas but teem in the country where the sky is pitch black, the turning of the planet is impossible to hear except where there is almost a total absence of any other noise to drown it out. Or so the story goes. It was an extraordinary explanation, and one I wanted to believe. Unfortunately, every scientist I've presented it to has said that there's not a kernel of truth in it. According to them, the rotation of the earth is a sound that simply can't be heard.

With some effort, I tugged myself from the jitterbugging grasp of these daydreams and refocused on why I had made this long pilgrimage to visit this ersatz lake. Through a series of political and environmental missteps, Cadiz Lake—or more precisely, the Cadiz Groundwater Storage and Dry Year Supply Program—has become a critical centerpiece of California's efforts to provide water to the nearly 23 million people who are expected to live in the L.A. metropolitan area by 2020. Put simply, the plan is to drill into the aquifers below Cadiz Lake and squeeze out any available water that's buried there. Wild estimates are being bandied about for exactly how much water flows underground, but no one has been able to produce hydrological evidence for any firm amount.

To trumpet the importance of the Cadiz undertaking, numerous briefings and press releases have been held, describing how it represents the most modern thinking in water management and calling it a model for public and private sector cooperation. Unlike most other prior water projects, we are told, a local company is taking on a portion of the financial responsibility and risk. Significant benefits are promised for the Los Angeles area, especially continued expansion of the region. I had made this journey to Cadiz Lake because I had to see for myself what all the commotion was about.

The full impact of what I found hit me only when I put Cadiz in historical context. The development of southern California is the result of what is without question the most aggressive and foolishly charted water management plan in the history of the world. A desert not fit for human habitation was blindly transformed, against all obvious environmental common sense, into a region of vast wealth, high consumption, and huge population. Now the state is so bereft of ideas for slaking the thirst of this water-starved, overbuilt area that it has turned to a dry lake in Death Valley where water never flows to bail it out.

The southern California experiment is not even a hundred years old, a minuscule percentage of the 4.5 billion years that the earth has existed. Yet it's already failing. All the original strategies for getting water into the region—from siphoning the Colorado and rural lakes hundreds of miles away to tapping the tiny Los Angeles River and its tributaries—are foundering, and officials are desperately trying to craft new ones.

What went wrong is simple. Nobody considered that aggressive water management—and creating a metropolitan region in a vast desert could be accomplished only with some of the most provocative water supply ideas ever conceived—has never been successful as a foundation for a region's development. It always ultimately backfires, resulting in natural disasters like floods, pollution, habitat destruction, or extreme water shortages. No one ever performed an analysis of how much water could reasonably be imported from outside sources to support people, industry, and farming without serious environmental ramifications—and thus, how much growth was actually possible—before southern California was allowed to grow unimpeded. As a result, after an existence that can be measured in mere decades, southern California is already running out of water. Its complicated network of long-distance supply sources is either depleted or providing far less water than before. Meanwhile, however, the region continues to grow at a rapid pace, as if no water shortage existed at all.

The Cadiz Lake program is an attempt to provide a new supply. But what it proves more than anything is that the water decisions being made in southern California are based on many of the same fruitless ideas that initially spurred the region's growth. Even more disconcerting, it illustrates clearly that L.A.'s options for resolving its worsening water crisis are narrowing. Standing before the canyonlike hole in the ground at Cadiz Lake, one thought kept running through my mind: *Didn't anybody visit this place?* There was no water; nothing was even wet here, and even an untrained eye could see that the prospects of that ever changing were few. The notion that somehow, through some over-the-top engineering feat and on the basis of wildly optimistic geological studies, this would become the latest oasis to slake southern California's thirst was, on the face of it, well, a mirage.

I felt compelled to make at least a symbolic gesture of help. I poured the water from one of the gallon-size antifreeze containers that Boyer had given me into the empty Cadiz basin. Most of it dried up in seconds, well before it reached the craggy lake bottom; it evaporated invisibly into the thick air. The thirsty earth immediately consumed the little that made it all the way down to the lake bed; in literally seconds, it was as if I hadn't even spilled any water at all.

The world's inhabited deserts are today in an extremely precarious position. With just a limited amount of local water to tap, they're profoundly dependent on outside sources for their supply. And with the demand for water increasing and the available amount dwindling, highly populated desert nations and communities are finding themselves increasingly defenseless. Just to import enough to survive, they are at the mercy of private companies and governments—which are themselves using water to strengthen their geopolitical position.

The Middle East, the Earth's most densely populated desert region, is a graphic example. Almost unnoticed by the rest of the world, which tends to focus on Israel's rift with the Palestinians and its other neighbors, some of the most complex and potentially dangerous jockeying for dominance in the Middle East is centered on control of the area's water supply.

Virtually every country in the region is running out of water quickly. With the region's population expected to increase by as much as 15 percent by 2025—to about 350 million people—its water needs will be double what they were in 1975. While the deepest-desert countries like Egypt, Syria, Iraq, Jordan, and Israel do have access to local internal water sources—so-called fossil water—these sources are being depleted rapidly and won't be replenished anytime soon. This water, found in underground aquifers, has been trapped since the last ice age; after it's used up, it could take thousands and thousands of years to restock.

It's therefore no surprise that projections for the amount of freshwater that Middle Eastern nations will have in the next couple of decades are spiraling downward. Jordan's water supply, for instance, has dropped from 906 cubic meters per person annually in 1955 to only 327 in 1990 and is expected to fall to 121 by 2025. Similarly, during the same period, Egypt's supply will have fallen from 2,561 cubic meters to 1,123 to 630. The same pattern is being played out in almost every Middle Eastern nation.

Since local aquifers are barely of significance any longer, three river systems must supply almost all the water for the region: the Jordan, the Nile, and the Tigris/Euphrates. Not surprisingly, bitter geopolitical fights over these rivers are already breaking out—some of which are recasting Middle Eastern political alliances in surprising ways—and bloody battles have already been waged over who has control of the water.

In the most striking incident, the Six Day War of June 1967 between Israel and Egypt, Jordan, and Syria, Israel quickly conquered the Sinai Peninsula, the Gaza Strip, the West Bank, and the Golan Heights. But while most people think of the war as yet another border dispute, it was actually fought over water. "People generally regard June 5, 1967, as the day the Six Day War began," says Israeli prime minister Ariel Sharon, who was a general in the conflict. "But in reality, it started two and a half years earlier, on the day Israel decided to act against the diversion of the Jordan."

Sharon's reference is to an Arab summit conference held in Amman, where Israel's neighbors voted to divert the headwaters of the Jordan—in effect, depriving Israel of its main water supply—in hopes that the nation would be weakened and unable to defend itself. According to Sharon, through diplomatic circles—often using the United States and the Soviet Union as go-betweens—Israel warned the Arab nations that if they choked off the Jordan, then Israel would retaliate forcefully. The Arab countries refused to back down, ultimately precipitating the war. Israel's eventual victory essentially gave the country jurisdiction over the Jordan. That shift in control is a pivotal reason why Jordan's late King Hussein, knowing that his nation desperately needed an assured, steady supply of water, finally agreed to hammer out a peace treaty with Israel in 1994.

Unlike the conflict over the Jordan River, no hand-to-hand fighting has yet broken out over the Nile, but Egypt's building of the Aswan High Dam (see Chapter 1), which traps the river in a vast reservoir and can divert much of its waters into planned Egyptian cropland, has created a nasty diplomatic rift. Ethiopia and Sudan, where the Nile's headlands are located, are concerned that with this project, Egypt is taking more than its fair share of the river. And they're worried that by stopping the Nile's natural course to the Mediterranean, Egypt is potentially reducing the amount of rainfall in the region and hence the replenishment of the Nile itself. At least publicly Egypt, generally a moderate nation, has said that it wants to work with Ethiopia and Sudan to achieve a satisfactory agreement over the use of the river. Still, Egypt isn't changing its plans to aggressively manage the flow of the Nile.

Even this struggle, however, is overshadowed by the most disquieting water conflict in the region, one that is more than anything a harbinger of how power in the world may eventually be commandeered by those nations who control the most water and can threaten to withhold it from countries desperate for it. That notion underlies the strategy being played out by Turkey, the only water-rich nation in the Middle East. In the past decade, under President Suleyman Demirel (not surprisingly a former water engineer), Turkey has laid claim to control the Tigris and Euphrates Rivers, the region's largest water supply, which emerge out of Turkey's eastern highlands. In the process, Demirel intimated that countries like Syria, Iraq, Israel, and Jordan should drop their long-standing enmity toward Turkey and support its positions as well as award water development contracts to its companies. Demirel also crushed a Kurdish rebellion and consolidated his power in the region. He is unambiguous about his approach. "The water resources are Turkey's; we have a right to do anything we like," he said. "We don't say we share the oil resources that other Arab nations have, and they cannot share our water resources."

This statement is particularly jarring, coming from the president of a nation that is 99 percent Muslim; the name for Islamic laws—*shari'a*—stems from a word meaning "the sharing of water." In the seventh century Islam emerged on some of the most barren landscape in the world—Arabia, Syria, Egypt, and North Africa. To spread the religion, Muslim leaders emphasized cooperation among their converts, realizing that with all of its enemies—Islam faced vicious attacks by Christians, Jews, and pagans—the new religion would be easily destroyed if infighting broke out among the faithful. And in that region, sharing what little water existed represented the very essence of cooperation. Consequently, Islamic laws codified the then somewhat revolutionary idea that all living beings have a right to water. One Islamic precept, for example, says that the people who dig a well may use it first, but they cannot deny water for drinking to either man or beast. Another precept holds that a man lowering a container into a well will have full possession of only the amount of water that fills it at that precise moment.

The Tigris and Euphrates, where the world's first civilizations of Sumeria and Babylonia were founded and where the biblical patriarch Abraham was born, pour out of rugged mountains near Mount Ararat, the spot where Noah's ark settled after the deluge. Both rivers, separated at some points by fewer than a hundred miles, travel south through Syria and into Iraq, past Baghdad, before emptying into the Persian Gulf. Since the early 1990s, Turkey has gone on a water project construction binge, earmarking $32 billion to build a series of twenty-two dams on the Tigris and Euphrates. These dams will let Turkey turn the rivers' spigots on and off, as well as divert their water at will.

Turkey demonstrated that it possessed this power to great effect in January 1990, when it stopped the flow of the Euphrates. The interruption took place, at least officially, in order to fill the vast lake that lay in front of the just-completed Ataturk Dam. But its real reason for shutting off the Euphrates was to demonstrate to Syria what might happen if it continued to aid the Kurdish separatists in southeastern Turkey.

Although Turkey turned the Euphrates back on three weeks later, Syria got the message. Its situation is particularly delicate. Syria relies on the Euphrates to irrigate its agriculture-heavy northern provinces, which form an especially critical part of the national economy. Forty percent of Syria's workforce is in the agricultural sector, amounting to 26 percent of the nation's gross domestic product. Consequently, Syria began to distance itself from the Kurds and curry favor with Demirel. It supported Demirel's threats against Iraq—mostly warnings that Iraq would lose access to the Tigris and Euphrates if it didn't drop its angry rhetoric against Turkey.

As for the Kurds, Syria's withdrawal of support for them, combined with the possibility that Demirel might cut off their water supply at any time, ended their attempt to create a separate nation-state within Turkey. They shelved their armed rebellion and negotiated a peace settlement that gave them few of their long-standing demands for autonomy.

Almost as soon as ground was broken on Turkey's first new dam, Demirel scored another diplomatic coup based on his water strategy: he forged a military and strategic alliance with Israeli prime minister Yitzhak Rabin, who was also a former water engineer. Under the agreement, Israel is guaranteed water from the Tigris and Euphrates. In return, Israel agreed to use some Turkish contractors for its water projects and also to back Demirel in Middle Eastern political initiatives.

Demirel may appear to have the upper hand now, but his is a risky gambit. Water is so precious in the Middle East—and water sharing there has been such a delicate balance, now set askew by Demirel—that defense experts are convinced that if Turkey ever followed through on its threats to withhold water from a particular nation, war would likely break out. Many Arab nations, even those whose relations are chilly now, like Syria and Iraq, could band together to take on Turkey, resulting in a bloody war over the region's water supply.

One leading Arab politician recently described the situation as follows, in a conversation that he was careful to say must never be attributed to him: "A time may well come when we have to calculate whether a war might be economically more rewarding than losing a drop of our water supplies."

Compared with the heightened tensions in the Middle East, the water crisis in southern California seems bland and innocent. But it's actually just as urgent. Look closely, and you realize that Southland (as the locals like to call the area) is essentially a theme park. Like Marvel's superhero island, where Spider Man and the Incredible Hulk have homes, or the space simulations and futuristic cityscapes in Epcot Center, the environment in southern California now bears no resemblance to the true geo-ecology of the region.

In Southland, the hidden landscape covered over by the ersatz rides and exhibits is a desert—and a relatively severe one at that. During an average year, fewer than ten inches of rain falls in Los Angeles. That's about one-quarter of the annual precipitation of a typical nondesert city, such as Boston or Brussels.

With limited precipitation and almost no underground wells or aquifers, Los Angeles should be an arid agricultural and ranching community, known for bean fields, orange groves, and cattle and sustaining not many more than the hundred thousand people—mostly Spaniards and Mexicans—who lived there in 1900. But the completion of the Santa Fe and Union Pacific rail lines in the late 1800s linked Los Angeles to the rest of the mainland, drawing tens of thousands of visitors to the city. Few had ever seen anything as limitless and innocently beautiful as the Pacific Ocean, and few had ever experienced such a perfect climate—so little rain, constant sunshine, and the pure warm air cooled by the water and the nearby mountains. The allure of Los Angeles had a mesmerizing effect, and wave after wave of dream seekers came to set up shop.

In the early 1900s, Los Angeles was home to an oil boom, a real estate bubble, a run on oranges, and a struggle for control of the docks. Starting in 1919, when Douglas Fairbanks and Mary Pickford purchased land and built their Pickfair estate, the motion picture industry migrated from the East Coast. Vast fortunes were made, and wealth accumulated more quickly than it had in any part of the country since New York emerged as the nation's financial center after the American Revolution. Los Angeles's political and economic elite arrived at a tacit understanding that no one was even to hint out loud at the possibility that the region's environment might be unable to support all this development.

Behind the scenes, though, the story was quite different. The city's infrastructure management teams, as today's urban planners would term them, were in constant crisis mode. The biggest issue they faced was water. Probably the best-kept secret (or the most-neglected obvious fact) in a city that even by the first decades of the twentieth century was driven by gossip was that Los Angeles circa 1920 was running out of what little water it had. The tiny Los Angeles River—in its natural state, for the majority of the year, not much more than a stream as it navigates southward from its headwaters in the San Fernando Valley to the Pacific Ocean in Long Beach, fifty miles away—was already sucked nearly dry. It was disturbing to those who noticed, but this tabescent strip of water was virtually the only local supply.

The region's water superintendent, a consummate deal broker named William Mulholland—mischaracterized as the effete, impotent engineer Hollis Mulray in the movie *Chinatown*—tried to stave off disaster by warning the public about the need for conservation. In one speech, he begged locals to consider water as precious as money. But mindful that growth was the city's first priority and was not to be interfered with, Mulholland stopped short of explicitly detailing the severity of the crisis. His public comments, weakly offered as they were, went unheeded.

At that point, as options were narrowing and the region was facing the likely fate of becoming a ghost village, Mulholland led one of the most drastic and aggressive water management coups ever attempted. To satisfy its dire need for water, the city commandeered the Owens River—a free-flowing basin more than two hundred miles away that drains melted snow from the Sierra Nevada and the Inyo range, on the Nevada border—and diverted it through pipelines to Los Angeles. In effect, Mulholland and his colleagues declared war on a distant part of the state and stole its water.

This action shook California from one end of the state to the other, and it has echoed since then, decade after decade, throughout the thin valleys of the region. This effort to supply Southland's water needs sundered the stability of California's environment so permanently that deaths, illnesses, litigation, ecological and engineering disasters, and unbending bitterness among counties—much of it still continuing today—can be tied directly to the Owens River coup.

That this exploit would have these consequences should have been obvious to Mulholland and the other L.A. water chiefs in the early 1900s had they carefully considered the situation. But they didn't—or more to the point, they couldn't. The decision to divert the Owens River was a thoughtless quick fix, to be sure, but it was also an imperative step if Los Angeles were to continue the wild growth that had inexplicably become its destiny.

Not surprisingly, precisely because the usurping of the Owens River was precipitous, done without any regard for environmental limitations or the natural course of water, it didn't resolve the obdurate problem that Mulholland and his cohorts faced: a desert, like Los Angeles, is by definition, literally, a place that isn't (or at least shouldn't be) inhabited. In other words, they either failed to comprehend or foolishly neglected to consider the fact that a desert's lack of water isn't merely an inconvenience that can be overridden by reverse engineering; rather, it is the defining aspect of the desert's existence. The absence of water is what characterizes the desert's evolution and restricts its ability to sustain life.

California today has many victims and unpleasant reminders of the Owens River debacle, but among the worst is Keeler, a town of fewer than a hundred people that lies in the southern dip of the Owens Valley. Although high mountains loom to its north, Keeler is mostly desert.

About the only going concern left in Keeler is the post office. And about the most energetic person in town, Geizel Rice, works there as the postmaster. Now in her forties, she came to Keeler twelve years ago from her native Costa Rica, drawn to the deserts of California. She still loves the area's dusky brown beauty, but Rice turns melancholy as she describes the drill of life in Keeler, in which toxic dust storms from the dry lake bed frequently blanket the community with potentially deadly chemicals, a direct result of L.A.'s Owens River diversion nearly ninety years ago.

"They come suddenly, dark clouds boiling off of the mountains and whipping through the town. There's very little warning, just a feeling that the breeze is picking up, and before long, you can't see more than a few feet before your face," Rice says. "By then, hopefully you've run for cover."

I don't think I'll ever quite understand how people who face such difficult and frequent tugs-of-war with nature can describe them so matter-of-factly, without much emotion, just resignation. But they often do. It's hard to comprehend the paralysis—often a personal decision, not even an economic one—that keeps people living in areas that are under environmental siege, their lives essentially on constant red alert. Even Rice couldn't clearly explain to me her decision to stay in Keeler; it perplexed her as well, she admitted. She could move if she wanted to. She can afford to and is mobile enough to do so, she says, "but this is my home. I really love it here."

Keeler's misfortune is that it is located on the east side of Lake Owens, which was carved out centuries ago by the Owens River as a huge natural drainage pit. The river needed the lake as a safety valve—a belt-loosening mechanism—to deposit some of its overflow. At one time, a hundred years or so ago, the lake was Keeler's reason for being. It was nearly as large as Lake Tahoe, and its deep blue water, reflecting the high desert vistas, attracted thousands of tourists. Captain J. M. Keeler, for whom the town was named, operated a steamboat on Lake Owens that ferried visitors and commercial traffic from one end of the valley to the other.

It's hard to engage anyone in Keeler in a conversation about what life was like back then. The lake is dry now—it's essentially a foe and is viewed with trepidation. It has been that way since soon after Los Angeles opened the spigot on its Owens River pipeline and flushed the region's water toward the Pacific Ocean.

The opening of the spigot was a defining moment for L.A., in many ways the moment that saved the city from simply drying up. Mulholland himself proudly topped off the event. Over the roar of the water that was now pouring out of the aqueduct, he yelled loudly to the large assembled crowd, "There it is. Take it."

No one in Owens Valley could hear this machismo, but they felt its effects. Even as Mulholland spoke, two hundred miles away, Lake Owens began to empty. There were a few reasons for this: Owens River didn't need the lake as an overflow basin anymore, because all of the river water was being driven west to Los Angeles by the diversion; and as the enormous pressure in the pipeline drained more and more water from the river's basin, the lakebed connected to the river was sucked out as well. Once it was barren, Lake Owens became an exposed dusty pit whose bed was laden with dangerous toxic metals—carcinogens such as nickel, cadmium, and arsenic, as well as salt, iron, calcium, potassium, sulfur, aluminum, and magnesium—that had previously been covered by the waters of the lake.

As mountain winds cut through the Owens Valley from over the Sierra or out of Nevada, they draw tiny particles out of Lake Owens's unprotected basin—specks so small that several thousand of them could fit onto the period at the end of a sentence. The resulting squalls of dust generate the worst particle air pollution in the nation. During these incidents, which occur as many as thirty days a year, the air pollution in Keeler is fifty times as bad as the worst Los Angeles smog.

The enormity of the windstorms' medical toll is painful to consider. On a per capita basis, the incidence of lung damage, respiratory disease, premature death, asthma, heart conditions, and cancer in Keeler and surrounding rural towns like Ridgecrest, Olancha, Lone Pine, Independence, and Big Pine, are among the highest in the United States. Personnel at the China Lake Naval Air Station as well as members of the Lone Pine Paiute Shoshone, Fort Independence, and Bishop Indian tribes are also affected.

"If the wind comes off Mount Whitney [the tallest peak in the state, located only a few miles from Keeler], we eat it," says Alice Robertson, who is raising two young girls in Keeler. "We can't see across the street. Afterward, I just go out and hose everything off."

That doesn't do much good, because as tiny as the particles are, for every one that lands on a birdbath, thousands have already lodged in people's lungs. It all happens very swiftly, according to those who have to deal most intimately with the aftermath of the dust storms month after month.

"When we see the white cloud headed down through the pass, the ER and doctors' offices fill up with people who suddenly got worse," says Bruce Parker, an emergency physician at Ridgecrest Community Hospital. "It's a pretty straightforward cause and effect. There's not much we can do for these people. We can alleviate their symptoms, but my God, we can't get rid of the dust that's in their lungs."

As disturbing as it is to talk to people in Keeler and witness the environmental cripple that it has become, it's fitting that the lasting outcome of Los Angeles's first foray into controlling water resulted in the destruction of Owens Valley's ecology and quality of life. What else could have been expected? Changing water's course is always a drastic remedy. The drying up of Owens Lake immediately after the Owens River diversion was a clear warning that such aggressive water management was foolhardy and would have only deleterious human and environmental consequences. Yet Southland's city managers refused to view it as a warning. Instead, they embraced this water management approach as a model.

The Owens River incident thus legitimized the idea that quenching Los Angeles's thirst for water was an acceptable priority, sanctioned by the state and even the federal government. And it could be carried out at any cost to the environment, habitat, or the public coffers and using any means.

That conclusion is clear from the way the city had obtained and held on to its rights to the Owens River. In the early 1900s, during Theodore Roosevelt's administration, the federal government had been exploring the possibility of using the river for irrigation projects to provide more water to the ranches and farms in Owens Valley. That exploration caught the attention of the L.A. water department, and Mulholland moved quickly to out-flank the government. He signed up J. B. Lippincott, the supervising engineer in Owens Valley for the federal Bureau of Reclamation, as a consulting engineer for the L.A. Water Department. With Lippincott in his pocket, Mulholland knew that he could stay one step ahead of Washington. Indeed, Lippincott's first move after being put on the L.A. payroll was to stop attempting to secure rights to land in the valley for the federal government.

Thereafter wealthy L.A. residents, led by former mayor Fred Eaton, began to purchase options on property in Owens Valley that would be needed for construction of an aqueduct. Eaton and the others, accompanied by Lippincott, told local ranchers and farmers that they were affiliated with the Bureau of Reclamation and that they needed the options to launch federal irrigation projects that would benefit the locals' operations. L.A. then issued a series of bonds to raise enough money to purchase these options and the underlying land at a premium from the power brokers that had acquired them. That done, Mulholland moved a construction crew in and began to build the pipeline. Despite the Lippincott double-cross, Roosevelt meekly acceded to L.A.'s plan, calling it "the greatest good for the greatest number of people."

Owens Valley residents were less placid in their response. In the early 1920s, as the aqueduct siphoned off much of their water, leaving little available for local irrigation, ranchers and farmers suffered severely diminishing output from their acreage. In retaliation, they seized the aqueduct gates and repeatedly dynamited the pipeline to L.A.

Mulholland, outraged, demanded that L.A.'s rights to the property in Owens Valley be safeguarded. A posse, backed by the state militia, poured into the area. In sometimes-violent skirmishes, the troops put down the revolt and thereby protected the pipeline. To this day, wounds caused by that civil war over water between these two distant California regions have not been salved. The amount of useable land in Owens Valley has shrunk considerably, and the area's residents accept only bitterly that L.A. controls practically the entire region: the city pays 80 percent of the county's taxes.

"Los Angeles owns this place," says Greg James, the water director of Inyo County, where Owens River lies. "There's little in the way of economic or public activity that doesn't have to go through the city of Los Angeles for approval."

If the Owens Valley residents failed to defeat Mulholland, however, nature succeeded. In March 1928, just a year after the protests were put down, the San Francisquito Dam—part of the aqueduct project, located north of Los Angeles in Santa Clarita—burst, unable to hold back the billion gallons of water in the reservoir behind it. A 180-foot-high wall of water plunged down San Francisquito Canyon. It took the floodwaters five and a half hours to reach the Pacific Ocean near Los Angeles. By that time, 470 people were killed, in the second-worst disaster in California history. (The worst was the 1906 San Francisco earthquake.)

These people weren't Owens Valley residents—faceless, far away, salt of the earth—these were L.A.'s own. Some were very rich and influential; others had left behind the pedestrian grime of the Rust Belt and industrial East to reinvent themselves in the Pacific dreamscape. In the eyes of many, they deserved better than drowning in an engineering mistake, killed by some miscalculation about water pressure. In the view of Southlanders, expressed loudly in the days following the San Francisquito's breach, someone had to pay for sullying the flawless picture that L.A. strove to present to the rest of the world. That person, of course, was William Mulholland, who was forced to resign in disgrace in 1928; like Owens Valley, he was a victim of his water policy. He died eight years later, a humbled symbol of failure, neglected on the fringes of the city's politics.

But Mulholland's water strategy wasn't discredited; on the contrary, his idea of acquiring water by whatever means possible, ignoring natural limitations, became a consistent thread running through the growth of L.A. In the decades after the city annexed the Owens River, Los Angeles water grandees built three additional pipelines: one from the Colorado River (year after year depleting more water from the Colorado than a 1931 agreement allowed, until a 1964 Supreme Court decision, ruling in favor of Arizona, put a halt to this); one from the Feather River in northern California; and one from a second source in Owens Valley, Mono Lake.

Each of these pipelines was necessary because after the Owens River diversion, Los Angeles found itself trapped in a vicious cycle. Seizing Owens River provided enough water for the people, irrigation, and industry that were in L.A. at the time. But that sufficiency itself encouraged more undisciplined growth, which led to the need for more water. And that meant more water pipelines and more importing of water from anyplace it could be taken. No one in power in the city government was willing to suggest that development should perhaps be moderated. And Mulholland's successors—though none ever obtained his kind of political muscle—dutifully followed in the path he charted.

This postponed the inevitable, which is only just now beginning to come into focus. Los Angeles, it seems, is having a much more difficult time keeping up with its vicious cycle. One problem is that as the number of people in the metropolitan area continues to skyrocket, the city's sources of water are not providing as much as they used to. The 1964 Supreme Court ruling required L.A. to share a greater amount of the Colorado River—from which the city gets most of its water—with Arizona and Nevada, whose own populations are increasing rapidly.

Even more significant, L.A.'s own environmental apathy is catching up with it. The sickness-inducing dust storms in the vicinity of the depleted Owens Lake as well as the draining of Mono Lake, to Owens's north—which occurred when the second L.A. pipeline in that region was built—resulted in several lawsuits brought by Owens Valley conservationists. In the mid-1990s, after twenty years of litigation, a state court ordered L.A. to fix the damage its policies had caused—including monitoring Mono Lake closely to make sure its water level didn't drop anymore and its animals and plants were protected. In addition, L.A. was told to rewet the dry Owens Lake bed. A small amount of Owens River water was sluiced back into Owens Lake through buried pipes and nozzles, creating a mud flap to hold down the dust, at a cost of $60 million.

L.A., however, can't afford to lose any more water. The cost of its defeat by Owens is potentially huge. In an action all too consistent with L.A.'s blindly drawn and crisis-driven water strategies, the city is now pressuring Inyo County, where Owens Valley is located, to let it make up for the water shortfall by boosting the amount of fresh groundwater that it pumps out of the region annually by nearly 50 percent, from 63,000 to 93,000 acre-feet.

Scientists exploring the region say that to increase L.A.'s subterranean pumping would be a bad idea, likely to bring more trouble. This action, says Wesley Danskin, a U.S. Geological Survey research hydrologist, would over time suck up the groundwater available to plants, deplete the wells that individuals use for their water supply, and create more dry lakebeds. "There's no free lunch," Danskin says. "The whole valley is interconnected."

In Owens Valley, L.A.'s request of Inyo County, occurring at a time when the city is being forced to restore the environment that it tragically harmed, is viewed as base arrogance, a sucker punch from a fighter who was just penalized for hitting below the belt. Still, because L.A. pays most of the county's taxes, it's unlikely that Owens Valley will be able to muster the political support to turn the city down.

Geizel Rice says she likes to think of her home as a small part of something much bigger and much more important than a speck of carcinogenic dust in a lake. Standing in front of the post office in Keeler, she pointed outward, toward the horizon, down the center of the rural road that cuts through town. The road had no visible end. On either side, dark fields clambered over small hillocks toward the mountains, miles away, torpid and unreachable. "This is infinity," she said. "Look around. It's an area where life seems to go on and on. There's something so wonderful about living where there are no limits."

Rice is frightened by L.A.'s actions and by its continued threat to the Owens Valley environment. She's perplexed by L.A.'s belligerent plan to siphon off more of the area's water supply—and likely cause additional long-term damage to Owens Valley—even as it has finally been forced to make up for the harm that its initial water forays into the region caused. "Just look at the record," she says. "When they bought the water rights a century ago, they didn't tell people they were going to dry out Owens Valley. But they did. How can we believe anything they tell us now?"

She turned to me and slowly measured her words. "Life may seem to go on forever here," she said, "but that's not really true. It only looks that way. It's an illusion. The real truth is, if our well goes, we go. Keeler would disappear."

Southern California's desperate search for new water sources has forced it to turn to the private sector for solutions, as have many municipalities around the world. Everywhere the worsening water equation is overwhelming the ability of governments and local authorities to provide creative solutions, while numerous for-profit companies have responded enthusiastically, suddenly seeing in water a commodity as potentially lucrative as oil. By necessity, the public sector is increasingly taking advantage of this private sector excitement for water, hoping to gain from the research and development that companies can do and from investments in joint water ventures that less cash-strapped corporations can offer.

There's a lot to be concerned about in this new public-private partnership. Companies are likely to neglect the environmental implications of their actions even more than are governments; the price of water may become so high as to be prohibitive for poorer countries, which will suffer more than they do now; and decisions about water development contracts could be guided by political cronyism, enriching companies and government officials at the expense of the local population.

Perhaps worst of all, the innovative ideas produced by the for-profits are at this point generally being plugged in to old bureaucracies, economic models, or water systems, which by and large are too sclerotic to absorb fresh technological solutions such as water transfer schemes, new storage techniques, and ocean desalination.

The locus of California's private sector partnership is Cadiz Lake, the putative future home of the Cadiz Groundwater Storage and Dry Year Supply Program and the desert basin where the only liquid I saw was the gallon of water I left behind in the open pit. The Cadiz project is the brainchild of a forty-two-year-old former investment banker named Keith Brackpool. In the past decade, Brackpool, a charismatic and well-connected Brit, has used a series of land acquisitions and influential relationships with top

California political figures to become one of the world's most active water entrepreneurs and a cheerleader in the state for the local private water industry.

Throughout much of the 1990s, borrowing a page from Mulholland's playbook, Brackpool, under the cover of his Santa Monica–based company Cadiz, Inc., quietly purchased about 27,000 acres of undeveloped desert. This huge tract, equivalent to the horizontal landmass of Switzerland, wasn't chosen idly. Cadiz's predecessor company was just coming out of a failed attempt to grow jojoba plants in the Arizona desert, and Brackpool's partner, veteran hydrologist Mark Liggett, suggested that the company's next venture should be to buy up inexpensive California acreage that has large water reserves under it, following the model that businesses use when they prospect for gold or other minerals. In time they could take advantage of L.A.'s increasing panic over water supplies and offer water to the city at a price that would pay Cadiz significantly more than the company had spent on the acquisition.

Brackpool and Liggett pored over dozens of satellite photographs taken of every corner of the California desert. The further away from civilization, the better they thought; the land would be easier to buy and relatively cheap. Finally, they found what they were looking for: images that appeared to reveal sinewy and serpentine underground drainage patterns beneath what is now Cadiz Lake, suggesting the existence of huge aquifers collecting water from a thirteen-hundred-square-mile desert basin.

Brackpool snapped the property up piece by piece, doing it slowly enough to avoid drawing attention to his activities. Though mostly close-mouthed about his business dealings, he told associates that he and Liggett were "mounting a water assets offensive. We're going to change the dynamics of how California views where water comes from."

In 1997, with the land fully in tow, Brackpool was ready to approach Los Angeles. Still a relative unknown in water circles, he booked an appointment with Woody Wodraska, then general manager of the Metropolitan Water District, the sprawling agency responsible for the water supply in twenty-seven southern California cities.

Brackpool's timing was perfect. Wodraska and Ron Gastelum, who in mid-1999 was named head of the Met, as it's known, fit the image of the new government water capitalists perfectly. In region after region around the world, water capitalists are replacing what were once known as water buffaloes, the slothful-minded but bullying bureaucrats who manage local water supplies.

Both Gastelum and Wodraska, who left California in late 1998 to become head of North American operations for Azurix, a Houston-based water company created by energy conglomerate Enron Corporation, are ardent water free marketeers. Hearing them speak, you could be listening to a CEO of a company selling toasters or big iron rather than a fragile, imperative natural resource. Gastelum, bald, with bookish eyes behind large-frame tortoiseshells, talks with some relish about his pre-MWD days, when he was in the private sector as general counsel to BKK Corporation, a hazardous waste and landfill company that has had a recent history of litigation and Environmental Protection Agency investigations over groundwater contamination at its sites. "I approach this job [at the Metropolitan Water District] the way I approached my job at BKK—that is, from the financial perspective," Gastelum says. "I'm here to provide a return to shareholders, to maximize shareholder values. I'm not here to write poetry. That means I have to do business by purchasing raw materials—in this case, water—at the best possible price with the least possible capital outlay." Wodraska's company, Azurix, a firm that Gastelum has been trying to work a deal with for water, is equally direct in its mandate: Challenge the status quo; create new markets; build the world's leading water company.

The water buffaloes, operating in quiet fiefdoms, out of view, nipping at the trough of government largesse, with little oversight of their monopoly activities, never had to use words like these or negotiate with companies that did. That's one reason they're now an endangered species; they didn't consider what would happen if their water holes began drying up, and they neglected to take into account that thirst has a price.

The tectonic shift from a slow, uncreative monopoly-driven approach to business and profit-minded attitudes is what Keith Brackpool, in one of those coincidences of luck and vision, walked into at his meeting with Wodraska in 1997. Brackpool showed Wodraska the satellite images of his desert property. By his reckoning, he said, upward of 150,000 acre-feet of water was sitting underground in the desert's aquifer, some of it fossil-aged, that could be pumped out each year and shipped to Los Angeles. (An acre-foot, or 326,000 gallons—approximately a foot of water spread out over a football field—is enough to supply a typical southern California family for two years.) He offered to sell the water to the Met for the exorbitant price of $230 an acre-foot. (Colorado River water, by contrast, because it is heavily subsidized by the state and federal governments, costs the MWD $25 an acre-foot.) Brackpool also said that in unusual wet years he could store as much as 500,000 acre-feet of excess Colorado River water on his desert property, in spreading ponds that would percolate underground. He proposed to sell that water, which he would save from flowing unused to the Pacific Ocean, back to Los Angeles at $90 an acre-foot when the Met needed it. Brackpool added one stipulation: the Met would have to pay at least half of the $150 million construction cost to build the pipelines, injectors, and sprinklers needed to bury and retrieve the water.

Wodraska was immediately taken by the plan and by Brackpool, even though the price was clearly high. For one thing, L.A. was running out of water, and any new, seemingly innovative ideas that had any chance of succeeding had to be considered strongly. Also, then Secretary of the Interior Bruce Babbitt was leaning on Wodraska to publicly support innovative and even radical water acquisition strategies for southern California. Doing so would provide Babbitt a rationale to tell Arizona and Nevada, which wanted more of the Colorado River for themselves, that California was doing all it could to develop additional supplies, but for now still needed the river.

"He told me that the pressure from the other states is going to be very intense, and that is why the Met has got to invest in the kinds of projects that show everybody how willing and committed we are to solving problems, [and how] we're willing to do more than anybody else in the West," Wodraska says. "It gives Babbitt cover to keep the aqueduct [to Los Angeles] full. It's a pretty cheap insurance policy."

But there was another equally important reason Wodraska was warm to the proposal. Although Brackpool was suggesting an ambitious undertaking without any proof that there was even water in the Cadiz desert or that it could store underground water, the prematurely white-haired Brackpool had an earnest, boyish elegance—a sense of class and purpose, not an unctuous pitch—that was unusual in the utility world. He represented a real business with a genuine business plan. Wodraska had felt for a long time that it would take just that kind of person to legitimize the water industry and come up with concrete answers, not just committee recommendations that never seemed to solve anything. "I had never run into anyone like that before in the water business," Wodraska says. "I wanted to make the deal work."

Wodraska had still another motivation for supporting the plan. After nineteen years at the South Florida Water Management District and nearly a decade at the MWD, he was ready to leave government and take a job in corporate America. He saw that Brackpool's idea could finally launch the private water business in California; it could be the pioneer that cut through a myriad of environmental and public relations obstacles. After Cadiz succeeded in at least creating a private water market in California, wherever Wodraska landed, he figured, he could use his contacts to take a cut of the significant water money that would no doubt be available. "I know of six other sites just like Cadiz," says Wodraska. "If Cadiz got off the ground, the others could easily follow."

Wodraska told Brackpool that he'd get provisional backing for the Cadiz plan, which would at least allow the environmental and geological analyses that were necessary before final approval to be done. But he warned Brackpool that coming up with $75 million in construction funds would require a bond offering, and that could take some time.

That was more than enough encouragement for Brackpool. On the heels of that meeting, with a preliminary deal in hand, he put in motion the second leg of his plan, one that had even longer odds of success than the first: to become California's top water policy maker, where he would be in a position to ensure that the bond offering passed and that it included enough money for his desert project.

To achieve this end, Brackpool, a horse-racing aficionado who owns stakes in Thoroughbreds, simply picked the right pony in the next important political race. Just months after his meeting with Wodraska, Brackpool went to work for the gubernatorial campaign of Democrat Gray Davis, a dark horse if ever there was one. Davis, the state's lieutenant governor who had been chief of staff under Jerry Brown, the colorful and enigmatically attractive governor in the late 1970s, was never thought to have a chance. His first name, it was often said, just about summed up his public persona.

But Brackpool felt that Davis—pro-choice, anti-guns, private to a fault about his Catholic faith—could succeed in the heavily Democratic state. In any case, Brackpool was certain he could have more impact on the soft-spoken Davis than on his higher-strung, more mercurial Republican opponent, attorney general Dan Lungren. And besides, Davis, while generally a loner, lived in L.A. and traveled in Brackpool's political circles. To Davis, Brackpool's Metropolitan Water District connection was a perfect calling card. So was Brackpool's money—he gave $134,000, either personally or through his firm, to Davis's campaign.

Unexpectedly, Davis won, buoyed by a flurry of last-minute donations and endorsements as well as voter fatigue toward the GOP and the two-term Republican incumbent governor Pete Wilson. Brackpool was re-warded almost immediately. Davis made him his chief water adviser and named him co-chairman of the Governor's Commission on Building for the 21st Century, where he worked on the resources committee, planning the state's water policies, among other things. Brackpool used his new sphere of influence—a significant one, in a state where water is among the most powerful political forces—to argue for the privatization of water at conferences up and down California as well as in individual and public meetings with state legislators.

He presented his message logically and compellingly, arguing, "Do the math. Is water worth more to a farmer for irrigation, or is it worth more as another crop that the farmer can sell to a city that needs it? Can an empty desert be put to profitable use as a storage zone to protect against water being wasted when it drains unused into the Pacific?"

Brackpool's ideas took hold among almost every special interest group in California. It was like a hypnotic spell. Remarkably, even California's ardent environmentalists were drawn to him.

That's hard to comprehend—until you dig deeper into California's water dilemma. It's gone on for so long and has been such an endless struggle among all regions of the state—each area is affected by L.A.'s enormous growth and need to import water from the north and west—that battle ennui has set in. In talking to dozens of people in California's environmental community, I got the sense that they've become too tired to argue about how to provide water to Southland anymore and that any new idea, even one that barely gives lip service to ecological concerns, is worth considering.

It's an unusually passive position for activists to be in, as Dan Sullivan, the Sierra Club's chairman of the state water committee, admitted to me. He tried to justify it this way: "You've been around here. Didn't you learn almost immediately that there's nothing surprising anymore in California water politics? Here's this private capitalist developer, who is the governor's confidant and is proposing a water policy that benefits him and may hurt the environment. And yet I sat through a meeting Brackpool headed up talking about water storage and transfers and marketing of water, and all of the environmentalists in the room were nodding their heads in agreement. You know what our position is—I know this will sound strange—but in this world if we don't put a price on water, then no one will treat it with respect. That's where Brackpool has a point."

Brackpool used the goodwill he gained in California to stitch together an alliance of upstate and downstate environmentalists, farmers, businessmen, and politicians to back a $1.97 billion bond offering that would, among other things, provide public funds for Brackpool's Cadiz desert project. Overwhelmingly approved by voters in March 2000, it was the largest water issue in California history and had something for everybody. Under its terms, $253 million was earmarked for watershed protection, $232 million for flood abatement, $345 million for clean water and water recycling, and $160 million for conservation. But by far the most money, $605 million, was set aside for water infrastructure and storage development projects, of which Cadiz was one.

Without this public money, the Metropolitan Water District would most likely have been unable to raise the funds to get the Cadiz project off the ground. That probability had been made clear just before California's legislature approved the bond measure and agreed to put it on the ballot. At that time, the measure stipulated that no more than $25 million could be spent on any one water project. But the Met's general manager, Ron Gastelum, told Brackpool that $25 million wasn't enough. To come up with the full $75 million necessary for construction at Cadiz, L.A. would need at least $50 million from the bond offering. Brackpool lobbied law-makers for a last-minute change that would double the amount of money that could be used for an individual water project and also drop the re-strictions on how much water could be extracted from an aquifer like Cadiz. The legislature gave in on both points.

"Unfortunately, we don't have the luxury of lining up a bunch of proj-ects and being able to just choose which one we want anymore," says Gastelum. "We can't be arrogant in the situation we're in. Cadiz may not be a panacea, but it's an extremely important piece of the puzzle. We couldn't afford to lose it because we didn't get enough money from the offering."

When the bond issue passed, L.A. began to take the final steps to close the deal on the Cadiz property. It was the first significant private water marketing program in California's history, and for Brackpool and his com-pany, it could mean a minimum of $500 million over ten years as well as significant juice to his reputation as an international water entrepreneur.

But one insoluble problem remains. According to scientists, there is at best a minimal amount of water in Cadiz.

A draft environmental review by the U.S. Geological Survey, released ten days before the bond offering, said that Cadiz's estimates contain "un-reasonable predictions of water levels and fluxes" and overestimates "the natural recharge to the basin by five to 25 times." According to the esti-mates, the moisture that will replenish the aquifers, which lie two hundred feet or more belowground, will come from precipitation drained from mountain ranges as far as forty miles away. That model, the USGS said, is "not defensible."

And then there's the issue of chromium 6. An unusually high amount of this element, which is suspected of causing cancer and other illnesses, was found in the Cadiz aquifers during tests. While chromium 6 can be reduced through filtration, this could add an expense to the project and perhaps slow it down.

"These signs are warnings," says Elden Hughes, one of the few environmentalists in California fighting against the Cadiz project. "They shouldn't be ignored."

Hughes is chairman of the Sierra Club's California/Nevada Desert Committee and a longtime foe of any development of the deserts. Unlike many of his colleagues throughout the state, he opposes water privatization generally, especially since the desert, the territory he's defending, has become one of the first targets of private water interests. It's generally thought that Hughes leaked to the press the Geological Survey report that debunked the inflated estimates of Cadiz water levels, well before the agency planned to release it. And he's been broadcasting the chromium studies, with obvious glee.

"I'm grateful for the chromium finding," Hughes says. "It shows that Brackpool will lie to get his way. Obviously, the water is not nearly as pristine as Keith Brackpool would have us think."

Hughes, who has been focusing on creating federal parkland in the Mojave and keeping landfills out of the desert, describes the Cadiz effort as L.A.'s latest land grab—"Owens Valley redux," in his words. But this one is laughable, he says. "At least when they took Owens Valley, they were stealing a free-flowing river. Now they're so desperate, they're raping a dry valley."

Hughes's comments, though, aren't likely to slow Brackpool down. L.A. is committed to moving ahead with the Cadiz project, and Brackpool is using that commitment and his relationship with Governor Davis to export his water business model to places far away from the state. In effect, California's efforts to solve its crushing water crisis have provided Brackpool with a pulpit to promote water privatization around the world and his own standing as a water entrepreneur.

Brackpool has been untiring in using this platform to his advantage. The most unsavory incident occurred in the fall of 1999, when Brackpool accompanied the California governor on a business development trip to the Middle East. There the pair met privately with Egyptian president

Hosni Mubarak. Out of that visit, Brackpool snatched a sweetheart deal to invest in one of the world's boldest water diversion schemes, designed to green 1.5 million acres of remote Egyptian savannah. The plan, an outgrowth of the damming of the Nile that created the massive Lake Nasser behind Aswan High, would transfer some of the water from this reservoir west to the barren town of Toshka. The goal is to transform about 100,000 desert acres into lush farmland and move as many as 3 million people—about 5 percent of the country's population—out of the downstream Nile Valley's narrow confines into this new thriving region. Mubarak, using the hyperbole that seems to frequently accompany water projects, said this diversion effort would establish "a new civilization" in Toshka to rival the one the pharaohs set up in the Nile Delta centuries ago.

To build the diversion project, Egypt will pay the billions of dollars, while Brackpool and other stakeholders, like international investor Saudi prince Al-Walid bin Talal, are now acquiring pieces of the Toshka property at cut-rate prices.

"You've got an excellent example here of what a public-private partnership should be," Brackpool said, on a visit to the site with cabinet ministers, bankers, farmers, and foreign diplomats.

Mubarak needs this project to succeed. The Aswan High Dam choked the Nile and cut into farm production. It led to an ongoing war of words, pitting Egypt against neighboring countries like Ethiopia, which claims that Egypt is using Aswan High to appropriate more Nile water than it has a right to under international agreements. The white elephant, as Aswan High is called in whispers in Egypt, has produced few obvious benefits for the country. Egypt continues to soar in population and poverty, and Mubarak and his allies desperately hope that the output from Toshka will feed the growing population and create jobs.

Some economists take issue with Brackpool's and Mubarak's assessment that projects like Toshka are smart economic policy, especially since they tamper with the course of water, a step steeped in unpredictable consequences. They argue that it makes more sense for a dry state like Egypt to import fruit and other crops—in other words, "virtual water"—than to spend billions of dollars to cultivate the desert for agriculture. The money spent on the water projects, they say, could be better used to feed, shelter, and clothe Egypt's poor population and, most important to provide water for drinking and hygiene in a country where many of its people go without.

Big efforts like Toshka "offer only a cosmetic solution, a way of not dealing with the real problem, which is represented in the daily struggles to survive of our people," an environmental consultant associated with the Egyptian Environmental Affairs Agency explained, with a deep degree of resignation about his country's future. He asked that his name not be used for fear of government reprisals. "Most people in Egypt," he told me, "will never see the fruit orchards in Toshka or probably ever gain an extra day of life from them."

Inhabited deserts like southern California give us a preview of what life would be like in a world with very little water—in other words, a world when the water crisis isn't solved. The desperate steps being taken here— searching for water in even the most fruitless places, diverting what little water is available and using every last ounce of it, even if it means immeasurably transforming the ecology of a region at the same time, forging dangerous geopolitical alliances to share scarce water—represent last gasps and foreshadow what the rest of the planet may be facing before long.

But most ludicrous of all, population growth in the inhabited deserts far outstrips that of the rest of the world. The world's deserts occupy nearly one-seventh of the Earth's land surface area, and most of them are unpopulated, as they should be—they have only four percent of the human population. Yet in the United States, the combined population of California, Nevada, and Arizona—the three states with the largest desert area—is expected to grow 62 percent by 2025. That's the largest projected increase of any region in the country. And every nation in the Middle East is experiencing a population boom that shows no sign of slowing down.

For the inhabited deserts, ignoring the seriousness of the water crisis—that is, turning to the same old solutions that have already failed and not limiting population growth—is a risky strategy. Many nondesert areas of the world are as water-scarce as the deserts are, and unfortunately more will become that way in the coming years. But the reasons for their water scarcity do not include a lack of water. The causes, rather, are a shortage of money, inadequate water infrastructure, the presence of pollution, poor water management, and government corruption. That set of causes, at least, can be viewed as a positive sign, since they are correctable. Some of the most water-starved countries have enough water—they've just failed to tap it appropriately or successfully. New technologies and fresh approaches to water delivery, combined with a global understanding of the water crisis and a concerted effort to resolve it, could pull these nations through. The inhabited deserts don't hold the same advantage.

3

Right or Need?

Water is life, and because we have no water, life is miserable.

KENYAN FARMER

We live hour to hour, wondering whether it will rain.

ETHIOPIAN HOMEMAKER

IN THE SEVENTH century, the laws and codes of conduct of the then fledgling Muslim religion unambiguously articulated that access to freshwater is a right of all living beings. The most precious of nature's gifts, water must be shared between man and beast, the Muslim rules of *shari'a* asserted. The idea that any person or group could control the availability of water or decide who gets water and who goes without was anathema and reprehensible to early Islam. This was a particularly enlightened perspective considering that Islam arose in the desperately water-poor Arabian landscape, where sharing water is perhaps the truest expression of altruism and at the same time the most difficult to live up to.

Today, some thirteen hundred years later, most of the world's regions face, with varying degrees of intensity, the same water scarcity as the Muslim homeland always has. But in the midst of our spiraling crisis, the governments and agencies that manage the world's water supply have responded with an official position that's much less progressive than the Islamic principles. In global environmental summits recently, those who hold the most sway in determining who gets water and who does not have concluded that access to water is, in fact, not a human right but only a human need like oil or transportation. Some local relief agencies and development organizations in emerging nations disagree with this premise, but those who hold the purse strings—and thus, the more dominant power in world political and economic circles—wholeheartedly endorse it. In the United States and other developed nations, this isn't an issue, obviously, because, true or not, there always seems to be enough water, and it is always available. But in most emerging nations, there's no certainty that the next glass of water will come out of the tap—if there's a tap at all.

The question of right versus need isn't a tiny semantic distinction or an intellectual argument with little significance in real life. The day-to-day living conditions endured by millions of people—and for some, their survival—as well as the growing gap between water haves and have-nots, are directly tied to the outcome of the debate over whether water is a right or a need. Moreover, global water management policies, many of which are so ill chosen that they're actually exacerbating the water crisis, are being crafted and funded based on its conclusions.

The clear-cut boundary between right and need may not be immediately obvious, but upon reflection the two words are found to be indeed strikingly opposite. A right is an entitlement; it cannot be denied without sanction. A need, by contrast, is something that is both necessary and desired but by no means guaranteed. If water is a right, then local governments are required to provide a clean supply of it to all citizens, but if it is a need, then they can classify water as something people want and perhaps even expect—like roads and telecommunications—but have no assurance of receiving. Simply put, if water is merely a *need,* then the existence of a universal human *right* to enough freshwater for an acceptable quality of life can be denied.

It's both perplexing and disheartening to realize that access to water has never been explicitly included in any of the various declarations of human rights that were produced by international agencies in the past fifty years. After all, no other natural element is so crucial to sustaining life, and the right to life is the most fundamental privilege that people have. The greatest oversight is in the extremely influential and detailed Universal Declaration of Human Rights (UDHR), which the United Nations General Assembly adopted in 1948. In the document's final wording, Article 25 states that "everyone has the right to a standard of living adequate for the health and well-being of himself and of his family, including food, clothing, housing, medical care and necessary social services." No mention of water appears there or anyplace else in the UDHR, even though the declaration specifically outlaws torture, slavery, and political persecution and declares that people have the right to work, rest, and leisure, among numerous other things. Leaving water out of the UDHR has had broad ramifications, because many emerging nations and international relief organizations have used the UN's human rights declaration as the outline for their own constitutions. Consequently, more people in the world are afforded the explicit right to join a trade union than to take a sip from a well.

An argument can be made, of course, that in Article 25, the United Nations meant to include water—and, indeed, did so tacitly, because fulfilling the promise of health and well-being depends on having access to water. But the minutes of the debate over the human rights declaration, as much as they are available, don't indicate this intention at all. The subject of access to water never came up in any substantial way. Perhaps better than anything else, this omission indicates how cavalierly the world has taken the possibility of global water shortages—and how our zeal for managing and controlling water has blinded us to the implications of not having water. Whatever the reason for the omission of the word *water* from the entire UN Declaration of Human Rights, however, its most damaging outcome, now that we face a serious crisis, is that it allows us to continue to neglect the problem and keep choosing water management options that are antithetical to resolving it.

By labeling water a need and not a right, economic and social development funding agencies can justify spending billions of dollars on massive but questionable water projects—mostly dams to produce hydroelectric power for industry—in poor African and Asian countries, while bearing no guilt for earmarking almost no money specifically to supply water to people's homes (except, ironically, in the form of floods).

The paradigmatic agency, in this respect, is the World Bank. Created in 1944 by the Allies to lend money to rebuild Europe after the war, it has refashioned itself over the years into the world's biggest backer of public works programs, which are ostensibly aimed at poverty reduction. The bank, which is based in the United States and is funded by the world's capital markets, has been particularly aggressive in its support for outsize water projects, allocating over the years more than $60 billion for dozens of expensive dams in poor countries. This expenditure has produced a spotty social record and a recent torrent of justifiable criticism. In return for World Bank loans—which often mean countries must put off paying for such things as education and environmental protection in order to cover the interest on this debt—the agency has generally dictated the kinds of projects that countries must undertake. In most cases, the bank has compelled the building of dams that favor industry and agribusiness. As a result, hundreds of thousands of people who lived near these sites, ironically without adequate water supplies of their own, have been forced to leave their homes and move hundreds of miles away to avoid being swept up in floods created by the new water project.

If I had any doubts about how crucial the distinction is between water as a right and as a need, they were eliminated when I met the residents of the village of Indanglasia, in Kenya. Isolated in a semiarid region just above the equator on the western edge of the country, no one in Indanglasia was aware of the heated argument taking place in more developed parts of the world over these words, but everyone in the town has felt the harsh effects of its conclusion.

In Indanglasia, securing water is women's work. More than once a day, wives and daughters trudge two miles or so over rock and caked mud to the Nzoyia River. There they fill jerry cans and jugs with as much as seventy pounds of water each and lug them back, often on their heads, up steep banks to their homes. It's considered beneath the men to pitch in.

At an early age, females in Indanglasia discover that life is severe. Childhood genital mutilation is practiced; girls have their front teeth knocked out as a rite of passage. And when they become women, they're not much more than chattel. Their nearly certain fate will be to marry a man who will have one or two other wives and to live cramped in a two-room thatched hut on a sugar farm with half a dozen children.

It's a life filled with constant irritations. During the day the incessant and inescapable sound of machetes cutting through sugar cane sometimes becomes so painfully unrelenting that just hearing it makes you feel as if your own skin is being sliced, one woman told me. The nights are much quieter, but not much better, she said. Amid the hushed shuffle of sugar reeds, the threatening hum of malaria-carrying mosquitoes keeps most people locked tightly in their homes.

Having to trek miles to fetch water from a river that is brown with muddy sediment for much of the year and caked with raw sewage in the dry seasons is seen as just another daily chore in Indanglasia. And it's viewed as more a necessity than a nuisance. Still, it's an extremely painful activity. That's clear from watching the body language of the women who walk the makeshift path, scraped smooth by their footsteps, to the river and back. Their legs trundle slowly, achingly so; their faces are without expression, and their eyes are vacant, as if being lost in no thoughts at all is better than facing what, in a rational world, would seem inconceivable.

I never heard them complain, though, because they have no reason to do so; transporting water is their responsibility. For the women of Indanglasia, the only way to get water into their homes is to carry it there over long distances. And doing that, they acknowledge, is decidedly their job. "How else is my family going to drink?" said Sundiata Kanana, a sallow-eyed, muscular 40-year-old whose deep black skin is streaked with white sugary dust from the air. "How else can I make the sick people in my house well and wash my babies?"

For a brief period of time, some years ago, the three hundred people of Indanglasia had hope that they could perhaps escape the full measure of this despair. Not surprisingly, it was Indanglasia's women who delivered this measure of optimism.

It began in the mid-1980s, when a UN relief agency that was working on bringing water to rural sections of underdeveloped nations offered the Indanglasians a deal. If they could raise 60,000 Kenyan shillings (about

$1,000), the United Nations would put up the other thousand dollars needed to buy pumps and other mechanical gear to tap Indanglasia's aquifers. Water still wouldn't be piped directly into people's homes, but it would be accessible within a hundred feet or so of each hut, instead of miles away. And since the pumped water would come, not from the surface of the Nzoyia River, but from captured underground pools, it would be cleaner. The only catch, besides the problem of coming up with the money, was that after the equipment was purchased, the Indanglasians would have to supply the labor to install the pumping apparatus, with guidance from the UN.

To Indanglasia's women, nothing could have led them to reject the offer. As the caretakers of the village, they understood, much more than the men did, how important water was to the survival of Indanglasia's families, how life there couldn't be sustained unless water became more easily available. The prospect of eliminating the long daily marches to the river and, as important, of having fresh water available nearby for cooking, drinking, and hygiene—probably more than they needed—was so desirable and so far beyond anything they ever thought possible that in their eyes no price was too steep to accomplish it.

A steering committee of about two dozen Indanglasian women was formed to oversee the water project. The first step, of course, would be to raise the money, a prospect that initially seemed incredibly daunting. In effect, the committee had to come up with more than twice the per capita gross domestic product of Kenya. To put that in perspective, consider how difficult it would be for a few hundred residents of one of the poorest communities in the United States—Harlan County, Kentucky, or Bedford Stuyvesant, Brooklyn, for instance—to raise $100,000, only from among themselves with no outside help or publicity.

Remarkably, the committee collected the money in a little more than twelve months. To do so, it imposed harsh, unprecedented financial measures on the village. Whole acres of future sugar yields were sold off to speculators who paid money up front for ungrown crops, at steep discounts to going rates. On top of that, the Indanglasian women essentially levied a tax on the earnings of the men in the community, requiring that a set percentage of their salaries be put into the pump fund.

"This was the most difficult thing to do," said one of the women who led the water project. Fearing consequences from the authorities, almost all of the participants in the water project refused to allow their identities to be disclosed. "It would be rare for men to work directly for women. But in this case, we had to take charge. And if they thought that meant we were their bosses now, we didn't care. The water mattered more than they did."

With the money raised and equipment purchased, construction began in early 1990. Again, few men participated. The women, though, did double duty, hauling water from the river for their homes in one trip and transporting even heavier bags of riverbed sand in the next. The sand was to be used to line and harden the trenches where the pipes, storage tanks, and pumps would be installed. These so-called bore holes were being dug simultaneously as women used picks and mattocks to chop through the cementlike, dehydrated earth.

Meanwhile, at the Nzoyia River itself, another group of women was building a subsurface dam, one that would lie under the waterline—it's not even visible from the banks of the river—and wedge sand behind it. Over time the captured sand would become the holding pen for an underground reservoir in the riverbed that resisted evaporation during the dry season. Once connected to the new pumps in the village, the reservoir would complement the aquifers as another reliable supply of year-round water.

The whole project was finished by 1991. Three pumps were fully installed, and pipes led into local wells and all the way down to the Nzoyia. The pumps barely coughed the first time they dripped water into the loamy dirt at the foot of the faucet, but a noise like a chorus—a gasp gutted by a sigh—rose from the residents of Indanglasia, all of whom had turned out to see water come out of the pipes for the first time.

After that, very little was the same in Indanglasia. Because the women were not tethered to the river anymore, they spent more time with their children at home, and many even took jobs, increasing the standard of living for their families with a second income. In addition, the water supply was soon chlorinated to kill bacteria and worms. That led to a marked reduction in infections like diarrhea and cholera as well as fewer cases of debilitating and sometimes fatal parasitic diseases such as bilharziasis.

The women who had participated in the project secured an additional benefit that was less tangible but noticeable nonetheless. Their status in the community changed. What these women had accomplished, in the eyes of the many devout Christians in that part of western Kenya, was akin to Moses getting water from a rock. It was on par with a miracle. No one expected the problem of water scarcity—and the grievous quality of life that came with it—to ever be resolved. And the women had achieved this while taking on Indanglasia's male-dominated structure. Through their actions and demands, they had forced the village's men to take orders from the women, if only for a short time. The combination of all of this triggered a slow but steady tilt in the way women were treated in Indanglasia. Nothing grand occurred quickly, but in the first steps, soon after the water arrived, women were at least seen as more than just the caretakers and caregivers; to a greater extent than before, they were taking on the role of equal partners. Major family decisions couldn't be made anymore without consulting the women in the house.

On a muggy day in the dry season of 1998, seven years after the pumps were first turned on, Indanglasia was suddenly stripped of its new water supply. No one knows how it happened or even when it happened, because no one heard a sound—or at least is willing to admit to it—but that morning the guts of the system disappeared. Thieves as quiet as cat burglars had slipped into the village during the night and stolen the pump heads and the mechanisms that bored deep into the ground to suck up the buried water in the subterranean reservoirs. All they left behind were useless sawed-off pipes sticking out of the ground like skinny tree stumps, connected to nothing.

After the discovery of the robbery, there was a ruckus and much posturing about catching the criminals, as would be expected, but nothing could really be done. The thieves were long gone, and they had left no clues as to where or who they were. The most likely explanation is that the pumps were stolen as part of a scheme to resell the equipment on the lucrative open market for new and used water parts. Prices for water infrastructure devices have been rising significantly as the water scarcity crisis worsens worldwide and more and more money is poured into upgrading water delivery systems, especially in developed nations. To feed this thirst, robberies like the one that occurred in Indanglasia are becoming routine. A suspicion persists in Indanglasia that the culprits may have been working for corrupt officials of the Kenyan government who have access to international water markets and, by selling these devices, could pocket quite a bit more money than they earn in public service.

In the wake of the theft, Indanglasia simply returned to what it had been like before the water system was installed. The brief improvements in health care and the economy lasted only as long as the pumps did. Now it's as if that short period of time never existed; the episode has been mostly expunged, at least from widespread discussion around the village.

There are some reminders left, though, especially among those who worked most passionately for the new water system—and who lost the most when it was stolen. One of these mementos stands in a tiny thatched hut in Indanglasia, leaning catercorner against the wall. It's the little metal pipe that the thieves didn't take, the discarded hull of the water pump. "When I saw what had happened, I knew that Indanglasia was going to suffer," said the woman who more than any other led the group that brought water to the village. "People may have forgotten what it was like before, but I hadn't, and I knew we were in for a painful and difficult time again. I thought to myself, while everybody was making such a fuss over the robbery, that I would take the little piece of pipe that was still in the ground. I put it against the wall and haven't touched it since.

"But my reason for wanting it wasn't so I could remember what we did here or what it was like when we had water piped near our homes. No, at the time I thought, foolishly I guess, that maybe someday we'd be able to use this pipe again. So I put it in this room, got my jug, and went down to the river to get some water."

The dusty, waterless thatched huts of Indanglasia, surrounded by sticky, mosquito-ridden sugar fields, are invisible to most of the world, hardly ever discussed or thought about—in contrast to the courts and palaces of the Hague in the Netherlands, where flower-strewn cascading fountains line elegant boulevards that are just minutes by car from the swank resorts of the North Sea. But Indanglasia and other desperate areas like it were much on the minds of the thousands of people who gathered in The Hague in March 2000, when a global environmental summit called the World Water Forum convened for a six-day meeting. At these sessions, attended by water management authorities, government leaders, water industry executives, environmentalists, and local relief groups—so-called NGOs, or nongovernmental organizations—the quarrel over whether access to water is a right or a need finally went public in an extremely rancorous fashion.

The WWF—an apt abbreviation—was called to bring together public and private sector representatives from every nation for a discussion of global water issues, including scarcity, toxic rivers, the deteriorating water delivery infrastructure, and potential hot spots, where skirmishes over water could break out. The organizers' hope was that the meeting would achieve a consensus agreement on resolving the worsening international water crisis.

It was immediately apparent that wouldn't happen. Private water companies used the meeting as an opportunity to do business, turned the halls of the WWF into a frenzied flea market. Shoved into tiny exhibitor booths, representatives of these companies screamed over one another to pitch their latest pipes, pumps, and water delivery schemes against backdrops of picture-postcard-quality shots of their most impressive projects. Deals were being cut on the floor, as water officials from developed and undeveloped countries pried instant discounts for construction and equipment from water sales reps, who peddled nonstop. Business got brisker and brisker.

Meanwhile, in the main meeting room, government leaders, international bankers, environmentalists, and NGOs were involved in a nasty argument that stretched over days. The World Water Council—sponsor of the WWF and led by international political and economic power brokers such as the vice-president of the World Bank, Ismail Serageldin, the former World Bank chief Robert McNamara, and the former Soviet Union president Mikhail Gorbachev—was presenting its "Vision" report. Billed as an overview of water crisis issues and possible solutions, this document was meant to be the first step toward producing a "Framework for Action," a plan to tackle worldwide water shortages, river cleanup, flood control, new water delivery technologies, and dam building, among all the other problems relating to water.

Most of the "Vision" report was uncontroversial. In sensible but actionless language, it called for such difficult-to-argue-with objectives as "empowering people, especially women, through a participatory process of water management; protecting ecosystems through sustainable water resources management; and cooperation in international basins." Many of the relief organizations at the session, however, were surprised by the directness of two succinct assertions that were almost buried among the document's thousands of words. One called water a "basic human need." The other said that water services should be priced "to reflect the cost of their provision."

This wasn't the first time these stances had been taken, but it was the first time those who have the most influence over the world's water supply set them down in writing so unambiguously as a statement of policy. Yet the debate over whether water is a right or a need had, in many people's minds, not yet been truly resolved. As such, for relief workers and water development organizations in emerging countries, the statement was a slap in the face. Not only did it define water as a need rather than a right, it proposed to base water management and delivery decisions on an acceptable return in dollars and cents—in other words, it embraced the idea that water is a commodity whose price depended on how much it cost to supply it—instead of taking the position that everybody must have equal access to water. NGOs saw these statements as a signal that big water projects—dams and hydroelectric plants—would, as before, have precedence over providing water to the world's thirsty. These organizations were desperate for money and had already suffered through decades of being ignored while water conditions worsened internationally. The water-as-commodity statements in the "Vision" report codified their worst fear that this situation would continue.

Somewhat surprisingly, a 1992 environmentalist-led meeting on water issues had initially legitimized the idea that managing water with an eye toward balancing costs against benefits is not only acceptable but even desirable and should be encouraged. Five hundred attendees representing a hundred countries had gathered in Dublin that year to prepare a list of water topics to be covered at the first Earth Summit, to be held in Rio de Janeiro a few months later. The upcoming Rio meeting, an extravaganza of presidents, prime ministers, and activists, had been called to handle a sudden global crisis that few seemed to anticipate. In nation after nation, environmental, economic, and social problems were converging, and together they constituted a threat to international stability and growth. Consequently, it was decided that they needed to be addressed simultaneously—with solutions that tackled all three areas—not separately anymore.

The sessions in Ireland, which were called the International Conference on Water and the Environment, offered four specific recommendations relating to water, which came to be known as the Dublin principles. The first three were, even by then, self-evident: Fresh water is a finite and vulnerable resource, essential to sustain life, development, and the environment; water development and management should be based on a participatory approach, involving users, planners, and policymakers on all levels; and women play a central part in the provision, management, and safeguarding of water.

The fourth Dublin principle, though, was a strongly worded statement, and coming as it did out of a highly visible meeting of environmentalists, it had a huge impact on the way the water issue has since come to be viewed. It stated: "Water has an economic value in all its competing uses and should be recognized as an economic good."

In similar form, that language was adopted at the Rio Earth Summit, and, as can be seen by what transpired at the World Water Forum in 2000, it has shaped the way the water crisis is now being addressed. Simply put, water has become a for-profit commodity. Businesses are emerging to buy and sell it, and water projects are now designed to ultimately provide a return on investment. The financial benefits of any given water deal matter more than the needs of individuals. To more radical environmentalists, the fourth Dublin principle was an unforgivable mistake. But most environmentalists—those who attended the Dublin meeting and those who didn't—continue to support it. In an unexpected twist, they've come to accept that the severe imbalance in water supply and demand requires that a price tag be put on water.

Typical is the view of Dan Sullivan, the Sierra Club committee chairman we met in Chapter 2, a twenty-year veteran of water conflicts in California. "When people view water as free, they don't conserve it, they pollute it, they perceive it as never running out," Sullivan told me. "But if it has a value, a price, they protect it. It's the difference between how we treat a diamond and a rock we find in the woods."

But water relief organizations in poor nations are concerned with issues much more fundamental than conservation; as they see it, you can't worry about conserving what you don't have in the first place. For that reason, the Hague water-as-commodity statements, echoing the Dublin principles, were particularly disappointing.

The NGO representatives at the 2000 World Water Forum at The Hague complained bitterly about the water-as-commodity clauses in the "Vision" report, calling them a surprise and a double-cross. During a series of angry shouting matches with WWF ministers, they argued that they had been duped into attending these sessions by the World Bank and its supporters, who were trying to claim global consensus for their free market water policies and hoping to give the impression that the NGOs supported those policies. The session was noisy, but the NGOs did not succeed in changing the outcome. The WWF ministers, not unexpectedly, unanimously endorsed the "Vision" report. They received overwhelming backing from the government leaders in attendance, even those from undeveloped nations, who relied on World Bank dollars to sustain their economies and felt they couldn't afford to alienate the agency.

Furious, the NGOs stalked out of the meeting and released a statement rejecting the WWF report. Water relief agencies from every part of the world signed the statement, including Kenyan Friends of Nomads, the Yemen Water Protection Society, the Council of Canadians, the Iran Center for Sustainable Development, and Nepal Water for Health. The NGOs called the WWF's "Framework for Action" a document that "emphasizes a corporate vision" and that gives "insufficient emphasis and recognition of the rights, knowledge and experience of local people and communities and the need to manage water in ways that protect natural ecosystems, the source of all water." And diving into the debate of whether water is a right or a need, they went on to say: "We strongly insist that a clean, healthy environment and access to basic water and sanitation are universal rights, and cannot therefore be negotiated as commodities."

In asserting this view, the NGOs were as frightened as they were angry. Uppermost in their minds was the dismal water management record of the World Bank—clearly the behind-the-scenes power at the WWF meeting and a significant influence behind the language of the "Vision" report. For all the billions of dollars that the bank has provided for water development projects—about 14 percent of its overall funding budget since its inception—most of the benefits have accrued to multinational construction companies and the largest local industries. Very little—well under one percent of the bank's spending—has been allocated for providing water to thirsty people.

The most vivid example of how such water profiteering has crushed individual rights, including the right to water, under its weight is the Narmada incident in India. In 1985 the World Bank approved a $450 million loan for the construction of the Sardar Sarovar Dam, a project that would create a giant reservoir and hydroelectric facility on the Narmada River in central India. By that time the bank had already funded nearly six hundred dams around the world. The Sardar Sarovar was just one part of an Indian plan to build more than three thousand dams on the Narmada and its tributaries, but it was among the biggest dams in the project. According to the World Bank, the dam would provide electricity for industry and supply water to as many as 40 million people in water-scarce Indian provinces. Completely unmentioned and entirely inexcusable was the fact that literally millions of people would have to be uprooted from their homes and moved into what are essentially refugee camps, at least temporarily, or they would be flooded by the lakes created by the Sardar Sarovar. Equally reprehensible, the money needed for the pipes and pumps to supply water to the thirsty in India, who were supposed to benefit from this project, was not even included in the funding.

The projected environmental damage that would be caused by Sardar Sarovar was just as bad. Thousands and thousands of acres of forest would be drowned, and the runoff would then pollute the rivers downstream. Consequently, the Indian water supply would actually be diminished by this project, not increased. Simply put, there were only two beneficiaries of Sardar Sarovar: the industries that would get cheap hydroelectricity, and the construction companies that would be contracted to build it.

Despite all of this—or more precisely, in ignorance of all of this, because the human and environmental impact was not even assessed before Sardar Sarovar was fashioned—the World Bank wholeheartedly endorsed the dam. Construction began in the early 1990s, and tens of thousands of Indian people, in what became known as the Black March, were indeed uprooted from their homes and land. In many cases, they were farmers and lost their livelihood along with their property. They were cramped into temporary shelters, waiting for what was known as rehabilitation, when they supposedly would get their own homes again. Many never got new homes, and today the water refugees, virtually penniless and with no prospects of jobs or income, still live in crowded halfway houses.

For nearly a decade starting in the mid-1980s, social activists and environmentalists lobbied U.S. lawmakers and World Bank members to stop Sardar Sarovar. They produced firsthand accounts of the forced displacements and refugee conditions that resulted from the construction of the dam, describing them as a calamity as bad as any civil war had produced. They offered evidence of the water pollution and deforestation that were occurring as well. Finally, under pressure from Congress, the World Bank agreed to conduct an independent study of Sardar Sarovar.

The bank's report, which came out in 1992, was stunningly forthright about the horrible human and environmental toll that the project had taken and the bank's complicity in it. According to the report, "The Sardar Sarovar Projects . . . are flawed, . . . resettlement and rehabilitation of all those displaced by the Projects is not possible . . . and . . . the environmental impacts of the Projects have not been properly considered. Moreover, we believe that the Bank shares responsibility with the borrower for the situation that has developed."

No one expected such a strong indictment of the dam or the bank, because the committee investigating the dam was headed up by Bradford Morse, a former Republican congressman who was a close friend of then World Bank president Barber Conable. But as committee members later recalled, it was impossible to disregard the fiasco that Sardar Sarovar was. Just by visiting the site and talking to people who were affected by the project, and by witnessing the suffering of the refugees and the potential scope of the environmental damage, the wrong-headedness of the project was obvious, they said. Perhaps worst of all, one of the members told me, no one at World Bank headquarters in Washington had even made a trip to India to see what the lending agency's money was being used for, despite many charges that this project was a mistake.

For a year after the report was issued, however, the World Bank continued its support for Sardar Sarovar. Finally in late 1993, unable to withstand growing public pressure from activists and congresspeople to end its participation, the bank finally told India that it was pulling out of the project and would not send the final loan installment of about $70 million. That, as it turned out, was only a small victory for Sardar Sarovar opponents. Even today the dams on the Narmada are still being built, using the initial outlays from the World Bank and left our funding from the Soviet Union before it broke apart. Tens of thousands of new refugees are created each year, and the water supply—in a country whose residents already have to survive on an average of only 31 liters of water per day—is both dirtier and diminished. And with the Indian economy sputtering—the country pays more money in interest to the World Bank for loans relating to Sardar Sarovar and other infrastructure projects than it takes in from taxes and investments—the scant additional hydropower produced by the Narmada dams is mostly going to waste; there's not enough new industry to support the potential increase in electricity. The big winners in the project so far are companies like the U.S.'s Enron, which received a guaranteed contract worth tens of millions of dollars to handle some of the larger hydropower concessions on the Narmada.

This episode, however, failed to convince the World Bank to reconsider its sponsorship of massive water projects. Instead of recognizing the senselessness and toll of such projects—or being concerned about the water crisis to which they contribute—the bank has still resisted funding local water supply efforts that would provide water perhaps to three hundred people (in Indanglasia, for instance), preferring instead to back projects with exaggerated and unachievable estimates of water for 40 million people. In fact, just a year after the World Bank pulled out of Sardar Sarovar, the lending agency gave China $670 million, its largest single loan of any kind ever, to build the Xiaolangdi Dam. This project, a five-hundred-foot high rock-filled hydroelectric dam on the Yellow River in the northern part of the country, will displace upward of half a million people. Even the World Bank admits that the Xiaolangdi project, which will inundate 186 square miles of some of China's most fertile land, is one of the largest forced resettlements of people it ever financed.

Costly debacles like Narmada could be avoided if we adopted the idea that access to water is an explicit human right. That's not to say that simply taking such a position would instantly remedy the acute water scarcities. Far from it. After all, the UN's Declaration of Human Rights establishes people's right to food, yet more than 800 million people on the planet are undernourished. Still, affirming that water is a right would be one step, important psychologically as much as physically, toward making access to clean water a global priority and responsibility, one that outweighs the construction of large expensive water projects with questionable benefits and undeniable drawbacks. More than anything, it would focus attention on many of the lamentable water management decisions that have been made—most especially, decisions that building a big dam takes precedence over supplying water to the thirsty. And it would create a specific national and international legal obligation for governments and world agencies to ultimately achieve the goal of supplying everybody with a minimum amount of fresh water—say, fifty liters per day.

It may not seem obvious to undeveloped countries that are burdened with immeasurable water scarcity and little money to do anything about it, but the economic and social costs of not treating clean water as a human right are actually higher than the costs of doing so would be. Without enough water, sickness and disease, frequently from birth, are endemic; the workforce is depleted as people die young; education is impossible for many because they aren't well enough to attend school or are simply too thirsty to leave their homes; malnutrition increases as crop irrigation founders, limiting the food supply; and people spend so much time trying to eke out even the smallest amount of water that it's impossible for them to be productive in any other way. Countries with scarce clean water are thus forced to apply their limited resources to shore up an overtaxed health care system and pay for barebones safety net programs that provide some food and shelter.

For a graphic illustration, it's illuminating to compare Kenya with Algeria, its neighbor to the northwest. The average Kenyan, including residents of the big cities like Nairobi, gets only about 36 liters of usable water per day. The situation is at its bleakest in rural areas, where fewer than half of the people have running water into or near their homes.

Algerian residents, by contrast, receive on average about 100 liters of water per day—but not because the country is overflowing with water. Quite the opposite: Algeria has just one major river, the Chelif, and it's only 450 miles long. But almost all Algerians live near the water. This wasn't the result of ingenious planning—it's more a natural imperative. Ninety percent of Algeria is covered by the virtually uninhabitable Sahara Desert, which is so hot and dry that it can barely sustain a thorny shrub. Only nomadic Bedouins have attempted to live there. Consequently, nearly all 32 million Algerians are packed into the Tell region, the hundred-mile strip in the northern part of the country along the Mediterranean coast. The free-flowing Chelif, which rises in the Atlas Mountains and widens into a powerful channel before reaching the ocean, easily supplies enough water for the people in the Tell.

Both Kenya and Algeria have potentially potent economies. Kenya has one of the strongest industrial bases in Africa, with factories that produce textiles and clothing, rubber products, transport equipment, and chemicals, and it has hundreds of thousands of acres of farmland. Algeria has widespread oil reserves and deposits of phosphate and iron ore. Both countries gained political independence—Algeria from France and Kenya from the United Kingdom—in the early 1960s. But that's where the similarities end. Kenya's water crisis has stultified its growth, leaving the nation a financial and physical basket case, while the steady flow of water pouring out of the Chelif has enabled Algeria to thrive.

In Kenya, with water-related diseases and unsanitary conditions rampant, the average life expectancy at birth is forty-seven years; in Algeria, it's seventy. Moreover, sixty infants out of every thousand die at birth in Kenya, while only forty do in Algeria. And Kenya's annual death rate is twice Algeria's.

With less water available to grow crops or support manufacturing facilities, as well as a labor pool enfeebled by the water crisis, Kenya's economy has fallen strikingly behind Algeria's. Algeria, for its part, has been able to invest its revenue in developing a rapidly improving health care system and an electricity, telecommunications, and resources infrastructure that supports industry and free markets. Consequently, Algeria's gross domestic product per capita is $1,600. Kenya's is one fourth of that.

Beyond these two countries, relief agencies and environmental analysts estimate that water-related diseases cost upward of $150 billion each year in medical expenses and lost work time around the world. By even the most conservative forecasts, about two-thirds of that amount annually would be enough money to improve water systems and provide new ones for regions that do have access to nearby water. That's still expensive, and of course, some of the water-related health care costs are incurred in areas without water nearby. But the large difference between the two figures at least indicates that in many regions it may be more expensive not to supply clean water than to provide it.

It's unfortunate that the issue of putting a price tag on water has been negatively linked with the question of whether water is a right or a need, because appropriate valuing of water is, in fact, critical to resolving the world's water crisis. The combination of viewing water as a universal right and pricing it so that it is used to cultivate global cooperation and conservation could be the centerpiece of a coherent strategy to confront the water crisis. In developed nations—even in barren deserts like California and Nevada—water is virtually free to consumers. In the United States, a country that believes fervently in free markets, such a low price for an item that is extremely limited in supply and high in demand is inexplicable. If people who can afford water were charged a suitable rate based on free market conditions, and if a portion of this money were used to subsidize water delivery and water system construction for less wealthy countries, the water market itself would remedy a good deal of the crisis. As water-starved countries slake their thirst on an increased supply, their standard of living should improve, ultimately enabling them to pay for their own water—and perhaps eventually subsidize other still thirsty nations.

Moreover, by pricing water internationally at a rate based on supply and demand in the richest free markets, private companies would have a profit incentive to develop advanced technologies for water distribution where water is scarce, such as water transfer techniques, desalination systems, and more efficient pumping equipment. The profit incentive is extremely important, because breakthroughs from private industry and for-profit science will also be essential to overcoming the water shortage.

Another advantage of appropriate pricing is that people will use less water. The Political Economy Research Center, a think tank in Bozeman, Montana, finds that when cities raise the price of water to individuals by 10 percent, water use goes down by as much as 12 percent. And when the price of agricultural water increases by 10 percent, usage goes down by 20 percent. Yet in the United States, almost no municipal water suppliers adjust prices seasonally, based on the actual supply of water available. Even worse, one-third offer quantity discounts to consumers who use *more* water.

More than a thousand years ago, when the Muslim code of water sharing was devised, the world was much smaller and less complex. The price of water, the policies of global lending agencies, the consequences of dams and planetwide water shortages weren't a concern. Now they are. Accordingly, our decision to not share water is akin to hiding our heads in the sand, a choice that the Muslims, who had more than enough sand and too little water, shrewdly avoided.

One of Islam's main worries was that without a universal right to water, people would go to war over it, and thereby rip apart the new Muslim society. With our mismanagement of water and our neglect of the water crisis, it's an outcome we may no longer be able to avoid.

4

Private Ties

WEEKS AFTER A months-long uprising had ended, blood was still visible on the streets of downtown Cochabamba, Bolivia. Most of the city's residents ignored the burgundy stains on the curbs and storefronts, as well as the shattered shop windows, though a few pointed them out to start conversations, telling heroic stories about friends who had fought in the conflict. Life had mostly returned to normal, which in Cochabamba is generally a poor, hard-bitten existence.

Yet the mood of the town had changed subtly. A distinct sense of pride hung in the air, which was out of character for Cochabamba, whose residents are normally quiet and fatalistic, reconciled to the unprepossessing qualities of their lives and not prone either to excitement over the good or to despair over the bad. But something remarkable had happened here. Cochabamba, an unknown city of a half a million people in the shadow of the Andes Mountains, had launched the first true warlike counterattack against globalization anywhere in the world.

The idea of globalization—essentially, borderless capitalism, in which worldwide economies are integrated and increasingly dominated by supranational corporations and markets—has certainly inspired dozens of street marches in many countries. The protesters' persistent criticism of globalization is that it opens up new trade and manufacturing markets for multinational companies without improving the financial condition of the people who live and work in these countries. Globalization's unspoken purpose, its opponents claim, is to give the world's superpowers economic hegemony over developing nations.

What happened in Cochabamba was philosophically linked to these arguments against globalization, but the incident was much more than just a political protest. It wasn't simply a political demonstration or a riot. It had the smell of war—and the bloodshed to go with it—and the unambiguous goal of a military campaign: to reclaim the city's water supply from Bechtel, a San Francisco–based multinational corporation that had taken control of it, after the regional government gave up trying to provide the city with a consistently clean and adequate daily amount.

It's not surprising that globalization faced its first serious challenge, its first physical battle, in a struggle over water. Who owns the water in a community as well as how much clean water is delivered to residents and at what price are questions that skirt razor close to the issue of survival—unlike, say, whether there are enough bananas, cotton, or even oil. This is especially true in a country like Bolivia, where water is at a premium. Each Bolivian resident receives on average 41 liters of water per day, 20 percent below the bare minimum needed for subsistence. Yet those numbers are deceiving. The largest group of Bolivians, the nation's poor—like those in Cochabamba—in fact get much less water than the average. The consequences for Bolivia have been grave: the lack of clean water is responsible for hundreds of deaths and illnesses there each year. In 1992 during an outbreak of cholera, five hundred people died in Cochabamba alone in a matter of months. That cholera epidemic eventually raged throughout Latin America and infected as many as a half a million people. It was bacteria in contaminated water supplies that spread the disease.

Embittered over these unrelenting conditions—and fearful that the amount of daily usable water in Cochabamba could diminish further under Bechtel, because the multinational might price water out of the reach of the poorer residents—the Cochabambinos mobilized themselves to go hand to fist with Bechtel in an effort to throw the company out. During these battles in Cochabamba, the commoditization of water finally collided with the idea that human beings have a right to it.

A violent face-off against globalization of water was bound to happen sooner or later somewhere in the world. A complex series of water management decisions and new water management realities—not just in Bolivia and some dating back decades—had made it all but inevitable. In perhaps the most important of these changes in how water was viewed, starting in the late 1990s, a handful of conglomerates began to quietly acquire control of the world's water systems. As the value of water began to soar along with the need for it, multibillion-dollar firms like Vivendi Universal, Suez Lyonnaise des Eaux, Enron, and Bechtel, among others, scoured the continents to, in some cases, purchase local water operations outright and in others to convince authorities to privatize their water concessions and let the companies run them.

It was a textbook example of globalization—in this case, of the water industry—symbolized by large water corporations opening up new international revenue streams, frequently but not exclusively in developing countries. And it spurred one of the most vigorous merger-and-acquisition markets in many years, as large water companies seeking to extend their international reach acquired smaller ones at a rapid clip. Between 1994 and 1998 there were 139 water-related deals, with a total value of nearly $4 billion. For the water delivery business, which until the early 1990s had been dominated by a group of provincial, old-fashioned monopolies, that pace of financial activity would have been inconceivable. But it was just a prelude. In a period of six months in 1999, Vivendi bought U.S. Filter, a western U.S. water operator, for $6.2 billion, and Suez Lyonnaise purchased United Water Resources, an East Coast company, for $1 billion. Those transactions came right after Enron, an American energy company, paid $2.2 billion for the United Kingdom's Wessex Water. At the same time, electronic water auctions debuted on the Internet, at sites like water2water.com and waterrights.com, where individuals with excess water, such as farmers with irrigation contracts, could put their water rights up for sale to the highest private bidder. As a result of all of these deals and new ventures, hundreds of millions of people worldwide now depend on transnationals—companies whose headquarters are thousands of miles away—for their water supplies.

This new dependence is occurring in rich and poor countries alike. Between 1990 and 1997 there were nearly a hundred cases of nonlocal private companies taking over the water supplies in developing nations; in the preceding six years there had been only eight. In the United Kingdom, the Thatcher administration privatized most state-owned regional water utilities in 1989. Upward of 15 percent of U.S. water systems have gone private. Although it is only in its infancy, the private water sector in the United States generates more than $80 billion each year in revenue—four times Microsoft's sales. In emerging regions the numbers are hard to come by—an indication itself of how heated water globalization has become—but there the rate of privatization recently has sped up even more. Worldwide, according to the World Bank, private water industry revenue approached $800 billion in the year 2000.

That's not to say that the transfer of the world's water delivery systems from the public to the private sector has gone smoothly. The performance of many water companies, which tend to be focused more on profits than on public opinion, has been shoddy and often insensitive to local concerns. For instance, recently in Walkerton, Ontario, a rural Canadian town with a population of about five thousand, a private testing lab that had taken over operation of the town's central well failed for five days to tell local authorities that the water had been contaminated with *E. coli* bacteria. The company claims that under its contract with Walkerton, it wasn't required to report this finding and that it was too understaffed to provide the notification. By the time the company finally informed officials about the bacteria, however, residents were already stricken by the water. Over the next few weeks, seven people in Walkerton died from *E. coli* poisoning and two thousand took sick.

Another botched privatization effort was the 1993 agreement that gave Suez Lyonnaise and Vivendi a thirty-year concession to run Buenos Aires's water system. Expectations were high in the local community that the two companies would rebuild the water delivery infrastructure and clean up a badly polluted water supply—after all, Argentina and Buenos Aires had actually paid the companies tens of millions of dollars to manage the water system and allowed them to keep most of the revenue they got from ratepayers. But seven years after the contract was struck, Argentine regulators called the program a failure: "The main goals set at privatization have not been met, in terms of the raising of water quality standards or in expansion of the system." Now, according to regulators, the only alternative to continuing the insufficient arrangement with the companies was the "doomsday option of revoking the contract." That would leave Buenos Aires, which no longer has an internal structure to run its own water system, facing an even worse crisis than it does today.

Despite fiascoes like these, private companies continue to easily convince local governments to let them take over water systems. To achieve this goal, according to municipal water authorities, water privateers use an encircling technique. They prepare detailed reports and proposals that are steeped heavily with statistics and conclusions supporting the premise that municipal control has been a failure and promising to supply water for the local government at a much cheaper price. Then they curry favor with local politicians through contributions, dinners, and tickets to events like the Super Bowl or the World Cup. In European countries, where bribing local officials is more common than in the United States, money changes hands as well. Then they deliver the presentation.

"It's hard for local governments to turn these companies away—they're everywhere, with arms like [an] octopus," says Douglas MacDonald, executive director of the Massachusetts Water Resources Authority. "They're almost impossible to defeat because no matter how many times you beat them back, they work a different angle to make the deal happen. In the end, they bid a price that's better than the local public water companies can, and they inevitably win. The consumer and, in any significant depth, the environment aren't even considered in this discussion."

MacDonald has lived through this process firsthand. He's been trying to fend off Suez Lyonnaise, which wants to subcontract Boston's water system and other local utilities for a fixed price—that is, with a promise that the amount it charges the Massachusetts authorities to supply water won't vary year by year. But that means, says MacDonald, that if water delivery costs rise, the price is likely to be passed along to the consumer. And without an assurance that water rates won't increase uncontrollably for area residents, which private corporations generally refuse to provide, MacDonald will keep arguing against the deal. So far he's been able to convince Massachusetts lawmakers and water policy chiefs to reject Suez Lyonnaise's bid—barely. He says he's resigned to the fact that he's likely to eventually lose Boston's water system.

I sympathized with MacDonald. Unlike other water officials I met, he actually understands the complexities of the water crisis. Many water managers rapidly lapse into speaking about safe subjects like abstract rates of effluence and pressure on aging pipes—in effect, they're drawn to engineering minutiae, not to the water that they're responsible for. But MacDonald uses language that is far more prosaic and realistic, if at the same time, for an official, somewhat poetic. I heard him give a keynote speech before the New England Water Environment Association, a group of water authorities, in which he was supposed to discuss water management design in the twenty-first century. He was bored by the topic and even found it irrelevant, so much so that he turned with enthusiasm and passion to discussing water's place in our lives as a cultural and sustaining icon, its meditative effects, the way the string instruments in Beethoven's *Pastorale* symphony "evok[e] a dappled brook," and how "there are just a few drops, figuratively speaking, in the streams, rivers and lakes on which we all depend for our existence on this planet."

Later, MacDonald told me that he knew most of his audience was probably unmoved by his wide-ranging monologue. They had come to the meeting to hear about the latest advances in water pipe design and sewage treatment procedures, not about art and nature. But he feels it's desperately important that everyone, especially water managers, come to grips with the idea that this water crisis will take much more than technology to resolve. However, before that can happen, he says, people must recognize a simple but long-neglected fact. "We never grasped that when nature offers us something as precious as water, it will also be fragile," MacDonald told me. "In my view, the vulnerability of water—the one natural gift we can't live without—is the single defining instruction, the simplest rule of thumb, the closest thing to a Platonic truth that we have for managing this planet, and we're not even close to appreciating what that means."

MacDonald admits that his holistic views about water sound, well, "goofy" to his colleagues, but in Massachusetts they have led to a well-balanced approach to water management that has produced decent results. The state's water system is now considered one of the most efficient and, relatively speaking, cleanest in the country; it is as strong on conservation and environmental protection as it is on supplying water itself.

But on the whole, most water officials around the world, by any measure, have a woeful track record. Governments, in wealthy and poor countries alike, have mishandled water supplies so badly for such a long time that their mistakes, poor planning, and ill-conceived policies are a prime reason that the amount of available water today is dwindling. Almost everywhere, the water infrastructure—the sewage pipes, the pumps, and the quality control stations—is falling apart, mostly out of neglect, creating vast pools of wasted water. At the same time, unchecked pollution is overwhelming the systems' ability to keep the supply clean. The anecdotes I hear about the condition of the world's water systems are so startling, they border on the incomprehensible. Developing countries regularly buy cast-off pipes on the open market to build brand-new water systems. Many of these pipes are virtually useless: they were rotted out or broken open during water main explosions. Whole portions of St. Louis's water system predate the Civil War, and the city is short nearly $4 billion to cover basic repairs and meet regulatory requirements. In New York City, six hundred water main pipes break each year. Most of these pipes are made of brittle unlined cast iron, and some of them are more than a hundred years old. Much of Los Angeles's water grid dates back ninety years, to when William Mulholland ran the department. All told, the Environmental Protection Agency projects that the U.S. needs an additional $23 billion annually to keep its water systems from deteriorating to an unacceptable level by 2020. On a per capita basis, the same is true for France, the United Kingdom, Germany, and virtually every other country in Western Europe. In some parts of Eastern Europe, the amount of money required to build and fix water systems is more than the gross national product.

Facing such abysmal conditions, water bureaucracies are more paralyzed than ever. Traditionally, their job has been to keep system-maintenance and -upgrade costs low enough—essentially, by providing Band-Aids instead of needed repairs and improvements—that they can hold water prices down. But as the water crisis deepens and the actual amount of water available for people, agriculture, and industry diminishes, even the limited dollars that municipal water utilities have spent on repairs are more difficult to come up with. For instance, El Paso, Texas, is struggling to find enough money for a sewer system renovation that the desert city has planned for years. Simultaneously, the city's water managers have learned that they will need a budget increase of 60 percent over the next decade to buy enough water to cover El Paso's needs.

These agencies historically never considered that water would someday come at a premium—that providing enough water for their residents would become a challenge. And they never took into account that water, after all, is finite in quantity and that there's relatively less of it as the population grows. By ignoring these realities and at the same time neglecting the physical infrastructure of their systems, public water authorities have hastened the crisis. Now they're boxed into a corner. They're paying skyrocketing prices to obtain scarcer water, and even the minimum upkeep of their water systems is a luxury they can no longer afford. A deteriorating system only exacerbates the supply imbalance and propels the price of water even higher.

This conundrum may make public utilities uneasy, but it's music to the ears of private water companies—in fact, it makes their case for them. They argue that they're driven by free market rules and the need to make a profit. Consequently, unlike the public water companies, they stand to benefit from taking advantage of skewed supply and demand. Not only will they build up-to-date, latest-generation equipment to match the increasing water needs of people and businesses and efficiently deliver water to them, but they will put a realistic price tag on water, one that will more than cover their costs. In a time of more demand than supply, that price tag will be higher than before, which will cause consumers to experience sticker shock. The high price tag, private water companies contend, will thus discourage overuse and encourage water protection and conservation. In short, the result will be a virtuous circle: Supply will increase because of infrastructure improvements and creative water supply techniques, and demand will fall because of appropriate pricing. Eventually, that dynamic will even out the growing water imbalance, and prices will dip to match more stable market conditions. Although the initial stages of privatization have been less than perfect, water companies maintain that it still is the best hope for resolving the water crisis and that over time the balance brought by a free market in water will iron out the problems. Anyway, they argue, the only other choice is to continue to blindly follow the public utility model that caused the water crisis in the first place.

What happened in Cochabamba put a very human face on the serious flaws in the water privateers' argument. The esoteric, tidy, and seemingly harmless notion of a virtuous circle was completely lost on this Bolivian city's residents, who had to live with its real-world implications. Instead of experiencing a perfect water solution, the circle's centrifugal force sucked them into a spiral that made their already difficult lives nearly unbearable.

Cochabamba is tucked into central Bolivia at the very edge of the Andes on the Rocha River, about 120 miles southeast of the capital, La Paz. Its water supply has always come from a web of tiny crystalline streams pouring off the mountains. Until just a couple of decades ago, tin mining made up more than half of the city's gross domestic product, and it provided ample jobs, many of them with wages approaching middle class. But tin prices collapsed in the 1980s, taking Cochabamba down with them. Now in Cochabamba there's a brewery and a shoe factory and the Cristo de la Concordia, an immense statue of Jesus Christ that has become a minor tourist attraction, but little else that generates revenue and jobs.

As the city slipped backward and the tin entrepreneurs exited with much of the money that had hitherto supported Cochabamba, the water system quickly fell into disrepair. By the late 1990s, it was in such bad shape that 50 to 60 percent of Cochabamba's water supply was wasted, seeping out through rusted holes in unmaintained pipes before it reached anybody. Water quality was barely monitored. Thousands of liters a day of the scant available clean water was siphoned off under the counter at discount prices to wealthier residents who had paid off the local water authorities.

Facing such conditions, most of the residents of Cochabamba were actually buoyed by the news in mid-1999 that the Bolivian government had sold the city's public water system to a subsidiary of Bechtel, Aguas de Tunari, and a group of British-led investors. According to the terms of a forty-year privatization deal, Bechtel promised to pour millions of dollars into expansion and improvement of the water supply. Under this private corporate regime, the Cochabambinos hoped, the dismal water situation would improve.

What they didn't know was that Bechtel had negotiated a couple of sweetheart clauses in the contract. For one, Bechtel would be allowed to raise water rates each year to match the increase in the U.S. consumer price index. Additionally, the agreement guaranteed the company an average 16 percent annual return on its investment, but the Bolivian government was in no financial position to subsidize this profit. Both of these clauses, then, virtually assured that Cochabamba residents would have to pay a lot more for water than before. In January 2000, about half a year after Bechtel took over the city's water operations and the same week that Aguas de Tunari finally hung its shingle on the water facilities, the water rates were increased significantly. For some, the monthly bill was doubled; others would pay three times as much as they had formerly been charged.

Pricing intervention to balance supply and demand is a frequently used globalization stratagem, but the Cochabamba economy—and its people— couldn't support so aggressive a market adjustment. Consider Tanya Paredes, a mother of five who lives just outside of town in a two-bedroom clapboard shack; one constant task of her daily life is to sweep out the dust that blows in from the pampas, so that her children will not come down with respiratory illnesses. Water is a big concern for Paredes. Many times a day the water pressure in her shack is so weak when she turns on the faucet, the result is a house-shaking cough of the pipes and then a feeble dribble of brown liquid. Plans to cook a meal or bathe a child or even have a drink of water are frequently altered, decided by the uneven output of her water supply. Paredes, a knitter in a textile plant, looks tired; her dark eyes appear dull, almost lifeless, as if they're covered by a smoky gel. She is resigned to her difficult circumstances. "It's just the way it is," she says. "You learn to live with what God gives you, because otherwise you learn to die."

When Bechtel raised Cochabamba's water prices, people in Paredes's situation viewed it as an obscenity. "It was as if they spat on us after we had already been raped by what came before," she said. Her monthly water bill ballooned from $5 to $17. That's about 20 percent of her salary of less than $100 a month. The $12 increase, she says, equals the amount it costs to feed her family for a week and a half. "We have to have water, so to pay the bill," she says, "I would have to cut down on my expenses for food, clothes, medicine, and the other things I need to buy for my children."

As rate increases rippled through Cochabamba, raising ire and fear, activists led by local labor unions rallied the city's residents to fight back against the higher water prices and Bechtel. As the unions saw it, a victory in the water wars could be beneficial to their recruitment efforts at the workplace. It would show them to be organizations that fight for broad-based community issues, not just for narrow labor problems; such a change in perception, they hoped, would overcome the workers' fears of employer retaliation if they joined up. One winning argument, presented by the unions in order to rally residents to fight the water increases (which were by then widely viewed as an evil of privatization), was that while Cochabambinos, to afford water, have to choose among necessities and do without some of them, Bechtel executives in San Francisco will regularly—perhaps many times a week and without any financial strain—spend more than the Cochabamba monthly water bill on one meal in a Bay Area restaurant.

In mid-January 2000 Cochabambinos made an initial foray against the water price hikes: a four-day general strike that left the city at a standstill. As many as ten thousand Cochabambinos participated in a mostly peaceful march to the city's central plaza. They demanded that water price increases be rolled back and that Bechtel promise not to increase rates again. Municipal officials agreed to negotiate with protest leaders and come up with a solution by February 5.

When that day arrived, no agreement had been reached. The protesters—now called the Coordinator for the Defense of Water and Life, or La Coordinadora, a loose-knit coalition of workers, environmentalists, peasants, and even water district members—marched downtown again, this time armed with sticks, rocks, ice picks, barbed wire, and knives.

The Bolivian government declared the protest illegal and sent in more than a thousand troops to disperse the crowd. The skirmishes turned violent and lasted a couple of days. About two thousand La Coordinadora and soldiers were injured. Plastic bullets fired from carbon dioxide charges permanently blinded two teenagers.

The next couple of months were nothing short of intermittent civil war in Cochabamba. Daily battles broke out for a time, causing scores of injuries. They ended only when the government pledged to cut back water rates and rein Bechtel in. When those assurances weren't lived up to, the war over the town's water ensued again.

It was a ragtag bunch of civilians that took to the streets: people in wheelchairs, seventy-year-olds on crutches, Incas from mountain villages, labor activists, and urban youths. This kind of mobilization was unheard of in Cochabamba. Even when the powerful miners' union organized a March for Life in 1986 to protest the privatization of the mines, the activists, met by soldiers, had dispersed without resistance. That bolstered the Bolivian government to destroy the unions, sell the mines, and begin a period of political repression and privatization.

But the struggle over water represented something much more basic. It exemplified nostalgia for a time when Cochabamba had been primitive yet economically balanced enough to attempt to supply adequate resources, and at the same time defiance, however vain, of a world that had clearly careened out of the local residents' control. This schizophrenia—part desperation and part rage—was captured in the shivering, defeated tone of one old Inca, who was marching behind a man forty years younger than he, brandishing an ice pick. Plaintively and manically, as if he were begging the gods to remember, the Inca screamed at the top of his lungs over and over, "Water is sacred." With each repetition, his voice grew a little weaker.

The government could do little but lie, saying that it would give in to the demands of the people, even though much of the time it didn't intend to or could not. Cochabamba and Bolivia were virtually bankrupt, so the official coffers were too empty to maintain the water supply. Turning water management over to a private company had seemed to be the only option. But sticking to that choice, at least in Cochabamba, would mean continuous unrest.

The Bolivian government, caught in a bind, but thinking itself unable to do without Bechtel, in March made a critical decision to escalate the civil war—in effect, to destroy La Coordinadora. Curfews and martial law were put in effect. Several people who broke curfew were shot on sight. Residents were told that if they were suspected of attempting to sabotage the water supply or even of refusing to pay the rate increases (as many were already doing), they would be subject to prosecution and beatings. It isn't clear how many Cochabambinos were killed during this time, but it's likely that at least six residents were murdered. Tens of thousands were injured. During late-night sweeps, Bolivian officers took dozens of La Coordinadora leaders from their homes and transported them to a remote prison in the jungle. Those were the lucky ones; most of them eventually returned. According to the Cochabamba Permanent Assembly on Human Rights, an unaffiliated watchdog group, an equal number of water protesters were kidnapped and were never accounted for again.

During the midst of the Cochabamba water war, Oscar Olivera, a lathe operator in a shoe factory and the leader of La Coordinadora, met me in a secret venue. It was early April, and he was in hiding, afraid of being arrested for a second time. His first arrest had been just a few days earlier, during a bizarre but typical incident in the Cochabamba contretemps. Olivera and fourteen of his La Coordinadora colleagues were at government headquarters, negotiating with officials in talks that had been set up by the Roman Catholic archbishop. Suddenly, a group of army officers, weapons draped over every inch of their torsos, barged into the meeting room and seized La Coordinadora.

"It was a trap by the government to have us all together, so that we could be arrested," Olivera said.

As word of this incident spread, more Cochabamba residents than ever before poured into the streets, this time with such boiling emotion that they spared nothing in the way from attack. Storefronts were smashed, house windows were broken, and opposing army and police were pummeled with rocks and Molotov cocktails. The Catholic Church and the local civic leaders pleaded with the Bolivian government to release La Coordinadora. Realizing they had made a serious miscalculation and that the situation was rapidly deteriorating, the government acquiesced, freeing Olivera and the other La Coordinadora the next morning. By afternoon, the government had apparently caved in completely: the archbishop told Olivera that Cochabamba officials had ripped up the water contract with Bechtel.

Jubilant La Coordinadora leaders walked onto the third-floor balcony of their headquarters in downtown Cochabamba and announced the news to the thousands of Cochabambinos in the streets below. Bells pealed from the city's cathedral.

"We have arrived at the moment of an important economic victory," Olivera shouted. "We have proved that the water is ours—we the people own it. Not private companies or even the government."

It was a short-lived success. Within hours the Bolivian government reversed the Cochabamba decision and reinstated the Bechtel contract. At the same time, whatever was left of civil rights in Cochabamba was suspended: gatherings of more than four people were outlawed, and severe limits were imposed on press freedom. Cochabamba's local governor resigned, saying that he didn't want on his hands the blood that would result.

When I first met Olivera, perhaps more as a measure of my own embedded expectations than reality, he immediately reminded me of Cesar Chavez, the labor leader who had organized the migrant farm workers in the United States and whom I had marched with numerous times at protest rallies in the late 1960s. It wasn't just the sallow skin and the tossed thick hair; rather, his remarkable face had two expressions that appeared almost moment to moment—one a sad, blowsy, almost unkempt acceptance of the indifference that permeates the world, and the other a taut defiance that jutted with the possibility of doing something about it. Both men were also extremely soft-spoken.

Olivera, forty-five years old when the water privatization battle began, was then head of the Cochabamba Federation of Factory Workers. This struggle, he told me, was different from all of the others he had been involved in as a union leader. The country had already sold off many of its formerly state-owned operations—mines, telecommunications, airlines, railroads, electricity—to private companies, and all the average citizen got out of it were worse services, fewer jobs, and higher rates. According to Olivera, privatizing water was a continuation of this trend, but it also cut much deeper. "The only thing that we had left, the only thing that was still ours, was water and air. Then they took the water away, too. We couldn't let that happen."

There was one more thing, Olivera said, that I should try to understand, although doing so might be difficult coming from the United States. "A lot of people believe that the new economic model of globalization has turned everybody into people who are out only for themselves and don't care about each other," he said. "We are proving by protecting our water that we have higher human values. We understand that water is a shared right, and that right is not for sale."

While I saw his point and was attracted to his passion for it, I couldn't ignore the fact that before Bechtel arrived the Cochabamba water supply was already a mess. Water may have been affordable, but there wasn't enough clean water being delivered, and the infrastructure was in ruins, neglected and getting worse. Bechtel, it seemed to me, may have highlighted the problem by raising rates, but it didn't cause it. And its goal, at least in the company's view, was to fix it.

I asked Olivera if Cochabamba's woeful pre-Bechtel water situation, mirrored in poor regions all over the world, wasn't indeed a strong argument for privatization of the water supply. After all, if local governments are unable or too poor to provide enough clean water, perhaps globalization, with its access to capital and its profit motive, is the only answer.

Olivera smiled. "Globalization. That's a word you only whisper in places like Cochabamba and not in polite company. You'll be surprised to hear, probably, that I don't completely disagree with you. But globalization cannot mean that the only ones benefiting in making the world one market are the companies. The local people are a part of this relationship as well— we are the consumers of the products being sold or the workers who make and deliver them—and there have to be benefits for us also. Not just promised benefits later on, but immediate ones as well. Nobody, even us in Cochabamba, thought very hard about this before they tried to steal our water, for our benefit supposedly."

As Olivera and I were speaking, just blocks from Cochabamba center, Victor Hugo Daza, an unarmed seventeen-year-old, was shot in the face and killed by the Bolivian army. He had been part of a motley crowd of a few hundred water protesters near the downtown square. This incident marked the end of the city's water war. Sadness replaced truculence, and nobody seemed to have any more energy to fight. Daza's killing finally drove home the point that this was indeed a civil war over water and that it was getting out of hand. Cochabamba was too fragile—the thread of its society too frayed—to survive continued fighting.

On April 10, pressured by Cochabamba's remaining authorities and by international diplomats who had ignored the crisis until Daza's death, the Bolivian government canceled its deal with Bechtel. The water system was returned to Cochabamba, and an onerous national law that had led to the privatization of the water supply was overturned. Future water management and development decisions throughout the country were put back in the hands of local communities, with the requirement that residents be included in the discussions. In addition, Bolivia agreed to give financial compensation to the families of people killed and injured during the water war.

The war in Cochabamba didn't receive much international publicity. Most of the news media ignored it. A few left-wing advocacy groups like the Democracy Center, whose executive director, Jim Schultz, covered the story tirelessly with daily on-the-ground, reasonably unbiased reporting, tried to generate interest, but the story was too complex, laced with too many difficult abstractions, such as the very real but intricate implications of the water crisis and water globalization, for the mainstream press to grasp and tell.

What's more, the street fights in Cochabamba did nothing to arrest the spread of water privatization. As a policy tenet, practiced everywhere in the world, it continues unabated. The conclusion that water is an economic good or a commodity—overwhelmingly ratified at the 1992 International Conference on Water and the Environment in Dublin and then expanded at the World Water Forum in March 2000 at The Hague—continues to encourage local governments and aid agencies to promote privatization.

Indeed, privatization and commoditization have become so sanctioned as a fundamental solution to the water crisis that the International Monetary Fund, which provides loans to nations in trouble, frequently imposes water privatization as a condition in its lending agreements. The rationale for this is the IMF's belief that structural reforms of the local economy and social infrastructure—everything from cracking down on cronyism and kickbacks to delivering services more efficiently to residents—are essential for a country to improve its financial standing. Because companies are more skilled at managing a water supply than the local government—and are more motivated by hoped-for profits to succeed—water privatization, in the IMF's view, can improve the lives of a developing country's population. As a result of this assumption, recent multiyear IMF loans of upward of $50 million each to some of the poorest countries in the world like Tanzania, Benin, and Rwanda specifically require that local water systems be turned over to private corporate interests.

But because of Cochabamba—and most importantly, because of the realization by companies, governments, and global agencies that people are willing to fight and even die for water—water privatization proponents have become extremely careful to avoid the clumsy way that Bolivia and Bechtel handled the transfer of the water supply. Governments have begun to realize that while privatizing water systems may be a desirable course for maintaining water supplies at peak level and distributing cleaner water to more residents, they cannot abdicate their essential responsibility to protect the needs of their populations. They are beginning to recognize that simply to hand over water systems to private companies and then turn a blind eye to their activities, as occurred not only in Cochabamba, but in Buenos Aires, Ontario, and dozens of other places, is dangerous. Governments are learning that for water privatization to succeed, they have to oversee the operations of the water companies, demand certain standards of performance and environmental protection, and ensure that consumers are able to afford the price of water.

The United Kingdom provides an apt illustration of how this can work. In 1989 much of the British water supply, then in dismal shape, was privatized. The plan gave private companies carte blanche to run local water systems, with no oversight. To Tory Prime Minister Margaret Thatcher, it was a perfect supply-side solution—turning the power of the free market to the public good. But it failed miserably. Financial mismanagement of the water systems became rampant, water rates skyrocketed, company executives gave themselves generous compensation packages from the water contracts, and water quality deteriorated.

In 1999, with Labour's Tony Blair in power, the privatization plan was amended. While private companies would continue to manage the nation's water systems, the Office of Water Services (OFWAT) was instructed to oversee and regulate the companies. OFWAT imposed rate reductions of as much as 12 percent and mandated certain required infrastructure improvements if the companies wanted to continue to do business in the U.K. This hard line was a quick success. In 2001 an impressive 99.8 percent of drinking water samples in the U.K. passed rigorous pollution tests, compared with only about 85 percent in the mid-1990s; pipe leakage, which had wasted significant amounts of water, has been cut by a third since it reached peak levels in 1995; and companies have promised to invest $72 billion over the next five years to ensure that supply meets growing demand.

The British approach—strictly overseeing the free water market, while still leaving it sufficiently unrestricted that profit incentives will motivate private corporations—has opened a lot of eyes, offering a model for how governments can control the potentially negative impact of water privatization on their populations. Consequently, even some of the poorest regions of the world—parts of India, Mozambique, and Manila, for instance—are adding clauses to privatization contracts that limit rate hikes, set predetermined investment levels, and establish performance benchmarks. Companies must agree to these clauses before gaining control of water systems. In addition, inspired by the accord that ended the water war in Cochabamba, these and other countries require an open accounting of the companies' activities so that local citizens can be involved in regulating them. Developing nations are able to do this now because the World Bank and the IMF, which encouraged Bolivia to privatize, and thus have been accused of complicity in the Cochabamba incident, are backing the governments' demands as the appropriate course of privatization.

Not surprisingly, none of these adjustments are dissuading private companies from bidding on water systems around the world or slowing down water globalization. Water is a commodity that's too potentially lucrative for private companies to ignore. And this could be the true—and unwitting—consequence of Cochabamba. A war fought against water privatization, led by socialists and local community activists, may have been the seminal incident that will eventually enable globalization to fulfill its promise as a solution to the water crisis.

5

To the Highest Bidder

Recalls Socrates: I went down yesterday to Piraeus with Glaucon the son of Ariston that I might offer up my prayers to the goddess; and also because I wanted to see in what manner they would celebrate the festival, which was a new thing.

PLATO, *REPUBLIC*

I WAS STANDING on a dock in Piraeus, Greece, staring at an odd-looking vessel moving sluggishly on the Saronic Gulf. The vivid sun was hot and, reflected in the waves, dizzying if you gazed at it too long. But I couldn't take my eyes off the boat in the distance. Getting smaller but clearly visible by its distinctiveness, the aging tug was towing skinny, bright red polyurethane bags, each a third the size of a football field, filled with water, to supply the half-dozen parched islands nearby. The Saronic Islands are surrounded by the vast Aegean Sea—one of the lushest bodies of water in the world—but with hardly any rainfall and just a few meager rivers that have been mostly squeezed dry, they are virtually without any freshwater of their own. Consequently, additional supplies of water have traditionally been shuttled there from Piraeus in the hulls of barges, the same way oil is.

The scene was an illuminating and disturbing metaphor for the depth of the world's water crisis: using the oceans to haul freshwater from places that have a surplus to places that are thirsty. It seemed to me a desperate endeavor, enlisting water itself to extricate us from thousands of years of dangerous errors in which we used up, mismanaged, and polluted the water that nature initially provided. I was certain that this couldn't be the way it was planned when the planet emerged. We weren't supposed to end up trading water on the high seas, ferrying it from where it is to where it isn't anymore or to where it never was.

Yet looked at in purely economic terms—and from the vantage point of a water-scarce twenty-first century—the tug and its cargo made perfect sense. We can't easily undo the mistakes and the accidents of geography, nature, and fate that have left people in some parts of the world with no local water supply worth tapping or at least none that is financially feasible to access. Consequently, a thirsty country such as Kenya could find that it's cheaper to purchase bags of water from a company in Canada, where more than 20 percent of the world's freshwater resides, and import them by boat, than to spend billions on pipes to pump in water from a polluted river only a few hundred miles away. In other words, at their best, seagoing water transfers could stimulate global water redistribution, diminishing the gap between the water haves and have-nots.

In fact, that potential—and the significant prospect of lucrative water marketing deals—has recently launched a spate of water transfer ventures around the world, most of them just taking shape and not unlike the one I was watching in Greece. Among the most aggressive are plans to draw down meltwater from Greenland's mammoth glaciers and sell it to any nation that pays the price. These ice sheets, which predate both people and pollution, cover 700,000 square miles (an area nearly three times the size of Texas) and are over two miles thick. Altogether they blanket 85 percent of Greenland. Like several other projects in that country alone, Aquapolaris, a partnership of a bottled water company and a shipping firm, has received a license to transport thousands of tons of melted glacial water to, in the company's words, "meet the world's growing water shortage."

Less benign water transport programs are also under way. Turkey's water-obsessed president, Suleyman Demirel, is shipping water by bags to Turkish-occupied northern Cyprus, a parched island in the Mediterranean with no permanent rivers of its own. As with most of his other water initiatives, Demirel has a high-stakes geopolitical motive behind his Cypriot strategy. He is hoping to force Greek-occupied southern Cyprus, which urgently needs water for irrigation and residential use, to reunite with the northern part of the island under his terms and under his control in exchange for Turkish water shipments.

I visited Piraeus because its water bag shipping system, run by a company called Aquarius Water Trading and Transportation, was the first to fully perfect the technology and demonstrate its viability as a means of supplying water to dry regions. Piraeus is Greece's busiest seaport, just five miles from Athens, but it is also a quaint city, made even more charming by the fact that it has never shed its ancient island roots. Sculpted by water, its geological past is evident in the narrow, serpentine roads that were squeezed into Piraeus thousands of years ago, when a slim inlet of water separated the tiny Mediterranean village from the Greek mainland. Now reconnected with Greece, after the water retreated permanently back into the ocean over time as Earth evolved, Piraeus is still drawn to the sea. Its hillsides—which rise sharply, as if on stilts, and are covered by neoclassical mansions—hang over the Mediterranean in the shape of a bow, the way a tree bends toward the sun.

Blacktop paths incline down toward the harbor, past pine trees that grow almost up to the Mediterranean's waterline, and at the bottom Piraeus's streets open into a seaside array of outdoor cafés and tavernas. There hundreds of people—many of them sitting alone, well dressed and restless, and others in small groups and mostly white outfits—eat freshly caught seafood, drink ouzo, and stare absentmindedly out to the islands in the Saronic Gulf.

Despite the idleness, it's a noisy setting: waves shuttle and lap against the docks, and moored caïques and yachts groan loudly. At intervals, longshoremen drown out even this din, yelling at one another as they maneuver mechanical dollies and lifts to load and unload everything from Fabergé eggs to Mercedes-Benz automobiles.

Taking all of this in, including the volcanoes and classical temples visible on the nearby Saronic Islands, it's easy to imagine that, absent some of the cargo and the diesel-powered equipment, the port of Piraeus probably didn't look or sound much different in the fourth century B.C., when Socrates journeyed here for a festival.

Aquarius's chief executive, a former U.K.-based commodity trader named Jean Claude Chalmet, was drawn to Piraeus in the early 1990s when he learned that daily water shipments had been taking place for years between Piraeus and the Saronic Islands. At that time the water was transported by barge, not a particularly efficient or cost-effective technique. More than anything, it's extremely expensive, because it ties up for days huge boats that were built to carry much higher-priced cargo and commodities. As a result, this approach had previously proved suitable only for atypical places like the Saronic Islands, which need extra water mainly for their teeming posh tourist trade and can pass on the exorbitant cost to their well-heeled visitors.

Still, as Chalmet saw it, the Piraeus-to-Saronic water shipments were much more than extravagant barge hauls, suitable only for the wealthiest water-starved regions. They were the germ of an idea that could perhaps help lessen the world's water crisis. Their efficacy, especially the seamless way that the water was delivered to people in homes, businesses, and hotels—few on the Greek islands even knew that most of their supply was imported—was evidence that transporting water could be suitable for quenching the thirst of a water-scarce location. Which meant, Chalmet was convinced, that it could be an extremely profitable business anywhere in the world, for companies that came up with a strategy to make it cost effective. In other words, something besides barges.

His interest piqued, Chalmet gathered a group of energy industry veterans who were experienced in the economics, science, and management of complex commodity shipments and set out to design a transportation system that ultimately could crisscross the world with water deliveries. In the first leg of their project, they developed a polyurethane bag technology for shipping water at a cost that was close to its price on the open market. When they finished, their company, by then called Aquarius, approached

Greek authorities and in 1997 won the contract to supply the Saronic Islands with water from the mainland.

Aquarius's technique is highly methodical. A series of metallic pipelines carry freshwater from the Greek mainland to the edge of the Piraeus harbor and then to an offshore loading platform a few hundred yards away. There the water is pumped into a succession of tubular polyurethane bags, each about thirty yards deep, which are covered with ultraviolet coating and a polymer developed for automobile tires to keep the sun from penetrating. Combined, these linked bags can handle anywhere from 1,000 to 2,000 metric tons of water. (A metric ton is about 2,200 pounds.)

As the bags are filled, Aquarius's staff monitors an array of dials to balance the amount of water going into the sacks with the sea's conditions—its depth and climate—to ensure that the bags will float during the voyage. The aim is to make certain that the naturally buoyant freshwater in the bags will not be heavier than the surrounding salt water. Ideally, to avoid mishaps during the trip, the bags should lie just below the ocean's surface as they travel, and in order to unload the water easily at their destination, they should rise to float as they are discharged. After the appropriate water measurements are checked and approved, an interlocking row of mechanical arms and clamps, moving with the precise and sequential grace of a daddy longlegs's limbs, herds the bags slowly toward a tug, where they are secured for transport. The whole process takes a few hours. The trip itself between Piraeus and the Saronic Islands is generally shorter than the loading operation.

The design for Aquarius's water bags grew out of a mid-1950s invention called the dracone, conceived by Sir William Hawthorne, head of engineering at Cambridge University and one of the developers of the jet engine. The dracone was a massive bag—it held more than 3 million gallons of water—and was much too heavy to float. In the 1950s, there was little need for water shipments, so no one even attempted to see if the dracone could be useful for that. Instead, marketed by Dunlop Rubber, it was put to work primarily to carry oil away from damaged tankers or spills and to transport oil to war zones and other distant outposts.

Because of their experience in the oil business, Chalmet and his partners knew of the dracone. In constructing Aquarius's first bags five years or so before winning the Saronic Island contract, they were faithful to much of its design. But the immense bags, when filled with water, were so unwieldy and overweight that, for their maiden voyage in the early 1990s, it took Aquarius more than a day to hook them up to the tugboats. When they were finally under way, the submerged bags made the tugboat so difficult to control that it got caught in a squall that ripped the bags apart, spilling the water into the sea. That lesson learned, Aquarius went underground for a few years. When it emerged with its proposal for the Greek authorities, its water shipment model was based on the lighter, smaller bags it uses today.

The current increase in water shipments around the world would not be occurring if barges were the only option for transport. But advances in bag technology have significantly lowered the cost of transferring water by boat. In fact, prices have come down so fast that seaborne water transport is now competitive with most other water supply technologies, including traditional approaches like pipelines and experimental ones like desalination. The biggest economic advantage of bag technology over barges is that the bags don't need to be immediately unloaded. They can be left behind, like a floating pipeline, serving as a storage vessel for the freshwater, which can be discharged as needed. This allows the tug to drop its cargo and return to its homeport for additional deliveries. Using a tanker for a shipment is expensive enough, but tying it up while the water is being unloaded and stored—a process that can take hours if not days—adds a significant cost to the transaction.

Water pricing is an abstruse process, because of the rainbow of subsidies and cost formulas that are used, which vary throughout the world. For example, southern California's Metropolitan Water District can pay $500 or more for each acre-foot of water it receives from the Sacramento River and other northern California sources. But by the time the MWD sells this water to its member agencies in desert towns like Los Angeles and Palm Springs, the price may be discounted by more than 80 percent after support from federal and local governments as well as accounting activities like depreciation and amortization are factored in. But most water economists agree that in the current market the price of water—including capital expenditures to develop the supply and deliver it while excluding subsidies—is economical when it falls below $1,500 an acre-foot. Transporting water by bags already costs about $200 less than that, and it's expected to get quite a bit cheaper in the next few years, as the technology improves. By contrast, water shipped by barge can cost as much as $3,000 an acre-foot, and desalinated water currently costs about $2,000 an acre-foot.

The significance of this price factor shouldn't be overlooked. Obviously, the most desirable approach for supplying water is a relatively pollution-free and well-maintained local water system. But that's an ideal that fewer and fewer countries are able to achieve. At current prices, water bag shipments are an acceptable alternative for poorer nations that otherwise could not come up with the huge upfront capital expenditures to immediately provide clean water from local sources or even to upgrade existing systems that are in disrepair. In other words, water imports—especially if they're subsidized by international agencies such as the United Nations and the World Bank as well as by surcharges on water use in wealthier nations—could become a quick-fix option for the thirstiest nations, giving them a chance to improve their standards of living well enough that they could then concentrate on cleaning their rivers, encouraging conservation and respect for water, and constructing a permanent water supply system over time. For those countries without sufficient local water sources, bag shipments, linked with conservation and population control, could be a permanent solution.

Aquarius's CEO Jean Claude Chalmet predicts that by 2010 tugs with water bags traversing the oceans won't be a novelty anymore but instead as routine as an ExxonMobil barge. When I asked him about this prediction, he said he's begun to think that his timetable may even be a bit conservative, especially considering that a half dozen or so new companies have recently entered the water shipment business and that a growing number of countries want to do business with his company. Aquarius is negotiating with Middle Eastern nations to provide water not only for their citizens but also to support industrial projects, mostly related to oil wells and pipelines. The company is also in talks with a group of African countries that are trying to get international funding for a range of projects that would use bag imports to create instant water systems. And Chalmet said he's working with water entrepreneurs in North America who want to create a transport corridor to carry water from Alaska, which has 40 percent of the U.S. water supply, to the desert towns of southern California.

Chalmet described the tone of all of these negotiations as increasingly frenzied—the way people get, he said, when they realize there is an actual crisis at hand. Details of contracts are being agreed upon more quickly, because nobody wants to be cut out of the available supplies. The situation reminds him of the oil shortages and energy traumas that he dealt with as a trader during the past three decades. People in developed nations took oil for granted and then panicked when they suddenly realized it might be running out, or at least that they didn't have complete control anymore over their supply. I told Chalmet that from what he's saying, it sounds like in the twenty-first century, oil and water actually do mix.

"Well, at least there's almost no difference between the two anymore," Chalmet said. "Both are provided by nature, and both are not always naturally located where they have to be." Both, he added, are vital to life on the planet, and are becoming scarcer, two characteristics that any commodity salesman would die for: "If you have a product that's washing into the ocean, unused, and it's needed desperately someplace else, why not sell it rather than let it run into the sea and go to waste? Why not make money out of it?"

Many environmentalists hate the idea of transporting water around the world and selling it to the highest bidder. They argue that with so few major rivers—from the Nile, to the Ganges, to the Yellow, to the Colorado—ample enough even to reach the sea anymore (largely the outcome of dams and other diversion schemes), our mismanagement of water has already created environmental and supply crises. Diverting more water from the few vital freshwater reservoirs that still exist would only worsen a perilous situation, hastening the destruction of delicate ecosystems and local water supplies.

A focal point of this contention is the Great Lakes, which contain about 20 percent of the world's surface freshwater and as such are an obvious target as a source of water for overseas transfer. But the Great Lakes are now fighting an uphill battle against industrial pollution, agricultural runoff, and overuse. As many as a hundred species and thirty-one ecological communities in the lakes are at risk of extinction, according to the Nature Conservancy. Environmentalists say that if large quantities of Great Lakes water are allowed to be shipped overseas, the level of the lakes would be lowered so severely as to jeopardize their very survival. Toxics would begin to consume the shallow pools and strangle the essential plant life that trout, sturgeon, herring and pike use for nesting and feeding. If that happens, the pollutants might seep into the water systems of dozens of communities that rely on the Great Lakes for their supply.

Critics contend that water transfers would not only lower the levels in lakes and rivers, subjecting them to overwhelming amounts of toxic chemicals, but could also affect aquifers, so integral to local water systems. Aquifers depend in part on precipitation for replenishment and cleansing. Yet already upward of 50 percent of the world's aquifers are dangerously shallow or are becoming more and more contaminated with toxics. Environmentalists assert that in areas where freshwater is siphoned off for shipment to other parts of the globe, less water will be available for transpiration and thus less rain will occur to stock and purify aquifers. As a result, the amount of clean water in these aquifers could diminish to levels unable to sustain adequate local water supplies. In other words, to quench the thirst of people three thousand miles away, we could choke off the supply for the residents next door.

The most passionate champion of this argument is Maude Barlow, who heads up the hundred-thousand-member radical environmental group Council of Canadians. She has assembled her thoughts on this subject in a widely distributed pamphlet called "Blue Gold," a well-researched and tautly written diatribe warning of the dangers of water privatization and commoditization. Barlow views compromise on any issue as a defeat for environmentalists; halfway measures and partial solutions are not half good, she believes—they're as bad as taking no action at all. When it comes to water, it should not be for sale, under any circumstances. "Water is part of our soul," Barlow told me after one of the hundreds of speeches she gives each year bitterly attacking those who suggest that economic globalization is necessary to improve the conditions of people in poorer countries. She has the bearing of a stubborn grandmother: warm eyes, a sympathetic smile, and cold resolve. "The future of one of the earth's most vital resources cannot be determined by those who profit from its overuse and abuse."

It's not by accident that Barlow is Canadian. Canada has become ground zero for the debate over water bag transfers, and it is the country that stands to gain or lose the most (depending on who's speaking) from this new technology. Possessing by far the most freshwater of any nation but with only 32 million people, Canada is the Saudi Arabia of water. Even after discounting for icebergs and glaciers in the mostly uninhabited northern part of its territory, Canada still has per capita four times the renewable water of the United States. The sheer expanse of Canada's supply can be seen in the water barreling over Niagara Falls, in the freshwater lakes that saturate virtually every town in the country, and in the huge open-armed drainage areas up north, where rain and snow run off unimpeded and seemingly without end into the Arctic, Pacific, and Atlantic Oceans.

Canada and Saudi Arabia, however, look at their endowments in quite different ways. Saudi Arabia has profited handsomely from its good fortune in having the greatest reserves of a commodity that is at a premium and in demand by the rest of the world. Canada, by contrast, is struggling with its twist of fate.

Canada has so far resisted attempts to share its water with other nations, declaring that it doesn't want to become the world's reservoir. Some of this attitude emerges from the growing influence of the Inuit, or Eskimos, on the country's affairs and even its culture. As Canada has begun to mend its relationship with its indigenous population, the First Nations, a powerful Inuit coalition, has gained a larger voice in national affairs. Among this group's most cherished tenets is: "Our relationship to water is one of stewardship rather than one of ownership." This statement fits well with broader Canadian beliefs, which in many ways have been shaped by the country's geography. With so much land and water (Canada has the second largest landmass in the world) vastly overshadowing its tiny population, Canadians are living with the evidence of nature's supremacy and have consistently shown an understanding that, despite the great impact that humans have had on the Earth, people are ultimately subordinate to the planet's natural cycles. Consequently, Canada has generally been one of the most pro-environmental nations in the world.

In the past Canada's decision to avoid bulk water transfers, which it has held to for decades, went mostly unnoticed, because the need for water in many parts of the world had not yet reached crisis proportions. Just as important, prior to the recent improvements in water bag technology there was no economical and, arguably, environmentally safe way to ship water to other nations. Most of the projects that Canada historically rejected were massive and clearly foolish water management schemes that, looking back, seem so obviously ill conceived, it's hard to believe anyone took them seriously.

For instance, there was the Grand Canal, a $100 billion project devised in 1959 by Tom Kierans, a Newfoundland mining engineer. The plan was to use dikes and pumps powered by nuclear energy to redirect freshwater from Hudson Bay in Central Canada south to the Great Lakes. From there, the water would be funneled to the Missouri and Colorado Rivers and used for irrigation; it would also be piped into homes in the Midwest and California, both fast-growing areas of the country at the time. This wasn't just a crazy pipe dream. U.S. water management officials were so dead set on building the Grand Canal and the Canadian government was so opposed that for a few years in the late 1950s and early 1960s, it threatened to harm relations between the two countries; just the mention of the project exposed raw passion. The project died, though, when legal delays and government wrangling scared off investors, who didn't want to put money into a project that had turned risky.

Around the same time, another even larger project was also under consideration—one whose scope, in retrospect, seems even more absurd. Known as the North American Water and Power Alliance (NAWAPA), it proposed nothing less than a total resculpting of western North America's watershed. The most outsize engineering project ever conceived, it was the brainchild of engineers at the Pasadena-based construction giant Ralph M. Parsons Limited, which has since been involved in building Ronald Reagan National Airport, Goddard Space Center, and dozens of offshore oil projects and water delivery programs in Alaska's Prudhoe Bay. Under NAWAPA's proposed blueprint, about 15 percent of the water draining off of the northwestern region of North America—an area that collects one quarter of all the rain and snow showering the continent—would be channeled into the Rocky Mountain Trench, a river-cut, glacial-carved valley running along the British Columbia–Alberta border. The Trench would thus be transformed into a five-hundred-mile-long, ten-mile-wide reservoir that would feed 177 lakes and rivers in the United States and Mexico with millions of acre-feet of water.

To create this huge water grid, as complex as the electrical utility system that crisscrosses the continent, the Yukon, Skeena, Fraser, Peace, and Columbia Rivers in Canada and the United States would all have had to be dammed. Those would have been the largest rivers affected. But beyond that, NAWAPA would have required 240 additional dams and reservoirs, 112 water diversions, and 17 aqueducts and canals. The estimated price: $300 billion.

When the Grand Canal was scheduled, NAWAPA fell into disfavor as well. But it didn't completely die; it just went underground. NAWAPA was so popular in the western United States that the region's politicians revived it in the late 1980s. The U.S. economy was just getting over a period of severe inflation, and the West was experiencing a stifling drought. Fear was widespread that a worldwide depression could be in the offing if farming costs weren't trimmed back enough to hold down food prices. Consequently, the price of water for irrigation in the western states became a huge concern. If farmers had to pay more for water, they would almost certainly have to charge more for their crops. So NAWAPA was revived as a possible safety net, a way to provide a steady, cheap supply of water for western U.S. agriculture.

Canada continued to resist NAWAPA, but in 1992 the project received a huge boost when environmentalists meeting at the 1992 International Conference on Water and the Environment, held in Dublin before the Rio Earth Summit (see Chapter 3), concluded that water should be recognized as an economic good—in other words, a commodity that can be bought and sold. Its sale no longer had to be confined behind national borders. With that declaration, especially coming from environmentalists, it seemed inconceivable that Canada could continue to oppose water transfers and reject NAWAPA. But at that time the Clinton administration was seeking Canada's support for the North American Free Trade Agreement (NAFTA), and eventually it decided not to attempt to strong-arm Canada into accepting NAWAPA.

Ironically, NAFTA itself may endanger Canada's reluctance to become a one-nation OPEC of water. Under NAFTA, which took effect in January 1994, the United States, Canada, and Mexico cannot restrict or prohibit the export of goods among each other. Ever since the Rio Earth Summit, the definition of goods has included water. And many recent export-import treaties brokered by the World Trade Organization, which regulates trade between countries, specifically include water as one of the products available for sale between nations. To complicate matters, however, a NAFTA side agreement states that the pact "creates no rights to the natural water resources" of the three countries.

The legal question here is: Into which category does water fit—good or resource? The answer is anyone's guess at this point. Recent binding and unbinding statements from environmental summits, the World Bank, the World Trade Organization, and other trade groups have so blurred the issue that the courts will probably have to iron it out. Many trade experts feel that even if judges eventually decide that NAFTA specifically prohibits water sales among the three countries—that is, if they affirm the trade pact's natural resources provision, they'll still have no choice but to reject Canada's resistance to selling water. This is because one of NAFTA's most central provisions backs the right of companies to sue if they believe that their ability to trade freely is hampered in any way. That core idea of NAFTA, legal experts say, would likely take precedence over the natural resources rule.

In 1993, when Canadians learned that NAFTA did not unambiguously exclude water as a tradable good, their reaction was fierce. During hours of debate in Parliament, ministers attacked then Prime Minister Brian Mulroney viciously. Typical was the comment of Francis Jackson, who represented a district in British Columbia: "This is the Sale of Canada Act. We have become a storehouse, a reservoir for the United States." Wendy Holm, a prominent and well-respected resource economist, presented a frightening vision of Canada's future when she warned that because much of the world is "deficient in water resources, Canada will be increasingly pressured to develop its abundant resource for the export market."

Meanwhile, as Parliament argued about whether the government had or had not protected the country's water in agreeing to NAFTA, a handful of entrepreneurs were already signing provisional pacts with local governments to sell Canadian water around the world. The Canadian government, feeling squeezed but still unsure of how to proceed, knew that it had to do something or else water shipments would begin soon. So in late 1999 it took a half-hearted step that satisfied almost no one: instead of clearing up the ambiguity in NAFTA by adopting an outright ban on Canadian water sales, Foreign Minister Lloyd Axworthy declared a moratorium on the bulk removal of water from Canada's lakes, rivers, and streams. He asked the nation's ten provinces, which have authority over local bodies of water, to pass permanent legislation with identical language to the moratorium. "This accomplishes the same thing as a ban on the bulk export of water from Canada, which we're frankly probably not allowed to do under our international trade agreements," he said.

Axworthy won over no environmentalists with that statement. Maude Barlow attacked the government, but in one of the more perplexing turns in this by now tortured issue, she faulted Axworthy for viewing water as a natural resource rather than as a commodity. By avoiding the trade issue,

Barlow said—by failing to take a strong position that Canada simply will not accept that water is a good for sale under NAFTA—and by classifying water as an environmental concern, the government was leaving open the possibility that the courts could one day approve bulk water transfers out of Canada, even if the provinces banned them. In other words, the judges might make a business and trade decision that overruled Axworthy's environmental stance. "If any province allows the export of water for commercial purposes, all of the bans across the country will be put at risk, because then companies could sue for unfair treatment," Barlow said. "By not having the guts to deal with water as a trade matter and dealing with it only as an environmental one, Axworthy is actually giving away our rights to our own water."

In fact, only half of Canada's provinces have backed the ban so far. None, however, have yet approved water transfers. Still, it looks increasingly possible that Newfoundland will soon become the first to allow the shipment of water. If it does, the water will probably come from tiny crystalline Gisborne Lake, on the southern coast of the province. In December 1998 a local construction company, McCurdy Enterprises, filed an environmental impact statement with the Newfoundland government, asking for approval to take about 25 percent of Gisborne Lake's annual outflow of 100 million cubic meters and ship it to the Middle East and the United States, among other places, where McCurdy claims to have signed contracts. According to McCurdy's figures, the $24 million project would provide $20 million in annual royalties and taxes for the cash-poor Newfoundland government. A water bottling plant that McCurdy plans to open as well would create 150 jobs in Grand Le Pierre, a coastal community that was hit hard when a North Atlantic cod fishery closed in the mid-1990s.

Jerry White, McCurdy's president and largest backer, describes the company's proposal as a plan to ship about 130 million gallons of Gisborne Lake water every two weeks—enough water for a conservation-minded city of nearly 200,000. Moreover, there will be no harm to the environment, White asserts. "The impact on Gisborne Lake is nil," he told me. "We're going to remove one inch of water from the lake; one inch, that's all. And by the time the boat reaches Canada's two-hundred-mile territorial limit, the one inch of water will be refilled in the lake. It'll be like nothing has happened here."

There's a lot of support for McCurdy's project in Newfoundland. The province is in an economic vise grip, and for many of its towns, it's extremely difficult to turn away new sources of money. "We're going right down the drain now because all the young people are moving away to find work," says George Fizzard, the mayor of Grand Le Pierre, where White's water transfer project supposedly would add dozens of new jobs. "There's so much water up here, it's hard to figure out why we wouldn't sell it, if somebody's willing to pay for it."

Newfoundland premier Roger Grimes recently echoed that point of view, albeit in somewhat hyperbolic language for his poverty-wracked province. "If water is the commodity of this century," he said, "then those who possess it and control it could be in a position to control the world's economy." Exaggeration or not, that statement has made it almost a certainty that McCurdy's project or one like it will soon be launched.

Considering the long struggle that environmentalists have waged against selling Canadian water, this defeat would seem to be particularly unsettling for them. But it's increasingly difficult to characterize the environmentalist opposition to water transfers as anything more than misguided eco-conservatism, particularly when the amount of freshwater available for people is diminishing rapidly around the world and potential solutions have to be examined openly, not rejected without consideration. The first time I met Maude Barlow, she was escorting Oscar Olivera, the leader of the anti–water privatization movement in Cochabamba, Bolivia, to demonstrations against globalization. The stories he told at these meetings about the vile conditions of so many dying South Americans who have only enough water to wash maybe once a week, and the statistics that Barlow offers up in her speeches to dramatize the number of countries facing water stress and scarcity—chiefly to make the point that the environment is worsening and the poor are suffering because of it—are actually the most compelling arguments for transferring water from areas with surplus to places with none or little. Indeed, despite her indefatigable and sincere concern for the environment, Barlow is beset by an inconsistency: as a way of humanizing her agenda, she dramatically and compellingly illuminates the water-starved Cochabambinos' daily strain to stay alive, yet she lives in a country overflowing with water and opposes efforts to share it with Cochabamba or any other place in the world.

There's an irony to this position, because the only Canadian water that is being transferred around the world without opposition from environmentalists including Barlow is bottled water, much of it bound for the wealthiest countries. The United States, for instance, which ranks among the top bottled-water consumers, along with France, the United Kingdom, Germany, and Japan, spent nearly $6 billion on these products in 2000, about a 15 percent increase from the year before and about 20 percent of worldwide sales. Purchases of bottled water in Bolivia, meanwhile, were negligible. That's not surprising when you consider that even inexpensive brands such as Poland Spring and Deer Park, not to mention premium products such as Snowline Natural and Trinity, cost as much as gasoline. Yet environmentalists don't complain about bottled water, for even if the most aggressive forecasts for future sales are met, the amount of the world's water supply transferred for these products will still be insignificant—less than 1 percent of available freshwater.

Barlow's seemingly contradictory stance about water transfers becomes even more baffling if one examines the procedures being put in place to control and oversee these activities. Environmentalist concerns about water shipments, particularly the assertion that severely diminishing freshwater levels would adversely affect ecosystems and aquifers, are valid— only assuming that the extent of these shipments is unregulated. Precisely because of environmentalist fears, however, governments and local authorities are already drawing up plans to manage shipments, including strict provisions to monitor the amount of water removed and ensure that the water replenishing a lake or river is of the same quality as the water taken out. If the rules are followed, transferring water, as illogical and extraordinary as it might be in an ideal world, may be a mostly benign and necessary evil in a world where region after region is struggling to find enough water to supply its populations.

Typical of the types of strict regulations being drafted for water transfers are those contained in a report by the International Joint Commission, a U.S. and Canadian organization that helps protect and manage the waters shared by the two countries. This document stipulates how to regulate Great Lakes water transfers and lists a series of environmental concerns that must be addressed before a program can be approved. Among them: the cumulative impact of siphoning water on the environment and aquifers must be measured and found to be insignificant; conservation practices have to be implemented; no net loss of water over time, beyond the natural amount of water lost by the lake, may occur (in the case of the Great Lakes, that's about 5 percent a year); and all water taken out has to be returned in a condition that protects the quality of the lakes and doesn't introduce non-native species into the water.

When I got beyond the dry and legalistic language of this and other similar water transfer operational reports that are now being drawn up around the world, it occurred to me that in one important way we've made significant progress. Had these documents been written just ten years ago—or perhaps even five—officials would likely have been cavalier, dismissing or neglecting the implications of diverting water and mismanaging the resource, the way we have many times in the past with the result that we now must even consider shipping water from one place to another. I was encouraged to see that if we must transfer water to palliate our mistreatment of it, the procedures we're drafting are so cautious and restrained that, to ecosystems and water supplies, it should be as if no water were even removed at all. I posed this point to Maude Barlow, asking her if she recognized that, perhaps partly due to the efforts of environmentalists like her, authorities were at least being extremely careful about not letting water transfers turn into an eco-nightmare, another failure in our attempt to control water. Barlow dismissed the thought as a noxious compromise. "Capitulation of any sort, no matter how pretty the package smells, ultimately stinks," she told me.

I understood Barlow's argument—and I appreciated that many environmental compromises have indeed backfired, enough to drive cynicism. But in this case, I couldn't agree with her position. By tiptoeing gingerly into water transfers, I am convinced, we are demonstrating a refreshing and essential difference from our more typical aggressive water management efforts; we are proving that we have perhaps finally learned something from our past mistakes, even if we still make too many of them. Knowing this, I am certain, will at least bring me a small measure of comfort the next time I find myself on a dock in Piraeus, Kenya, Chile, or Santa Monica, staring at a tug pushing water bags on the sea.

6

The Scramble to Restore

There are no other Everglades in the world.

MARJORY STONEMAN DOUGLAS,
SUFFRAGETTE AND ENVIRONMENTALIST (1890–1998)

IT WAS CHRISTMAS Day 2000, warm in Florida as usual, when a reporter drove to Belle Glade, the biggest city on Lake Okeechobee, with Claro Rodriguez, a sixty-six-year-old retired sugar worker who came to the United States from Cuba in 1968. Okeechobee—in Seminole, it means "big waters"—is the United States' third-largest freshwater lake. Its runoff, over thousands of years, has created the Everglades, a stunning five-thousand-square-mile subtropical wilderness of bog and river grass that lies directly to its south and extends to the Gulf of Mexico. The roads Rodriguez traveled were worn and weathered, jutting with pockmarks and potholes. All around were lush Okeechobee farmlands, acres and acres of sugarcane, long rows of rice and beans—and the abject poverty and aimlessness of migrant farmer existence. Makeshift shacks, their doors wide open, were crowded with people spending the holiday together, eating small portions on paper plates. Here and there groups of three or four people—but sometimes a lone individual—were sitting in a field, on a rock or tree stump, peering into the distance. His car was one of their few diversions, and they stared at it as it drove by, drinking it in from its front grill to its fin.

It was unspeakably sad. It was difficult to mask distress about the surroundings. "Does it look that bad to you?" Rodriguez asked. He sat straight and proud in his seat, his muscles still firm from years of hard labor, his clothes carefully ironed, his face clean-shaven, his wisps of gray hair combed back and oiled to his scalp with brilliantine. "Well, maybe yes, it does look bad. But people here have work and enough to get by. And the land around Okeechobee is very good, very, very good."

Downtown Belle Glade offers a different set of depressing conditions. Its suffering is commonly captured in one statistic: Ten years ago it had the distinction of being the U.S. city with the most AIDS cases per capita, chiefly the result of intravenous drug use. But even with that in mind, and steeled for the worst, it's impossible not to be taken aback by how dilapidated and run-down the city actually is. Its strip shopping centers and laundromats were all in disrepair; store signs were broken, thick tape was cut diagonally across dozens of windows, and old paint jobs were peeled like broken skin. The streets looked gray, even in the hot midday sunlight. Stray dogs freely roamed from one side of the road to the other. Rodriguez had to slow his car down to avoid hitting them. Otherwise, the town was empty and silent. Though it was Christmas, no one could be seen or heard celebrating.

The Okeechobee farmworkers may not have a lot, even by lower middle-class American standards, but considering the existence they fled from in their homelands in Cuba, Mexico, and other Latin American countries, they view it as bounty. Now they're worried that this new life is endangered by a plan to restore the Everglades to the way it was fifty years ago and in the process save the giant wetland from eventual extinction. Once the most complex ecosystem in North America, the Everglades has now dwindled to half its original size, and the water that once flowed freely through almost a third of Florida barely moves anymore. To restore the water flow—and, one hopes, the natural terrain—massive amounts of water from Okeechobee will have to be diverted both to the Everglades and to thirsty cities on Florida's east coast. As well, thousands of acres of local farmland will have to be flooded. In the midst of a severe drought, Lake Okeechobee, which irrigates the fields owned by the sugar companies for which Rodriguez and his friends work, is already dangerously low. The thought of siphoning off water from the lake for all of these purposes is terrifying, Rodriguez said.

By this time, Rodriguez had reached Okeechobee. For a huge lake, it seemed remarkably tame. Towered over and encircled by a worn flattop wall, it looked shallow, as if it were slipping deeper and deeper into its own bed. "I've never seen the lake this low," Rodriguez said. "I know something must be done, but I don't trust the government. If they try to take away our water or our land, we won't let them."

In its pristine state, the Everglades was one of nature's most spectacular and elegant water systems. The watershed flowed clockwise from Okeechobee to the coral reefs and estuaries of Florida Bay (which lies between the Florida Keys and the mainland), a sheet of water creeping almost imperceptibly through autochthonous grasses. The lake's clean water sustained life in the Everglades, which was the habitat for hundreds of species of snakes, snails, lizards, and birds. And with its natural pumping system, made from intricate layers of shallow pools, reeds, and marsh, the Everglades helped regulate the distribution of water throughout all of south Florida. Even now, to feel the pure wet ground underfoot; to see the tidal bays, lakes, and linked waterways that honeycomb the wilderness of tropical savanna and virgin forest; and to hear the cries of birds so distinctive as to seem otherworldly, mingled with the crunch of creeping but unseen animals in the bulrushes, it's difficult to believe that the Everglades is no longer as it has always been.

But in the late 1940s, to curtail random flooding and unlock more land for development, an elaborate plumbing scheme was constructed on Lake Okeechobee that diverted much of its runoff directly into Florida Bay. A little of the remaining excess continued to feed the Everglades, but most of it was delivered through manual pumps to cities in the east that needed water. This system was a death sentence for the Everglades. Choked off from a large portion of its water supply, fully half of the acreage that used to be vibrant wetland was either filled in and developed or left to stagnate and be overrun with weeds. Sixty-eight of the Everglades' plant and animal species are now on the verge of disappearing, including the manatee, the seaside sparrow, the wood stork, and the Florida panther. The number of wading birds—principally herons and egrets—has dropped by 90 percent. Exotic plants such as the Brazilian pepper and the Australian pine are suffocating native flora and altering habitats.

Now Florida is frantic to undo this damage. In a joint effort with the federal government, $7.8 billion has been earmarked to rebuild the Everglades, the largest environmental restoration project ever. More than sixty separate construction projects are involved, and it could take decades to complete. As ambitious as this makeover is, no one questions its necessity. Even people like Claro Rodriguez who fear its impact on themselves and their livelihood are convinced that the Everglades has to be restored. The main reason: Florida is rapidly running out of freshwater. With its aquifers dangerously depleted, Florida can't afford to lose forever such a massive, water-rich ecosystem. The state is in a race against time to turn the clock back, to a period when the marshes, swamps, and wetlands of the Everglades on their own, without human intervention, rhythmically and unfailingly distributed water through the area with the precision and consistency of a heartbeat.

Because it's one of only a few attempts to restore a vast, complex water management system, the Everglades project stands out among hundreds of other water renewal efforts that have sprung up recently, mostly in developed nations. But the others, which more often than not target aging dams, are equally important. In the past few years, upward of five hundred dams in the United States have been destroyed or been slated for the scrap heap, because over time their environmental costs—harm to aquatic ecosystems, poisoned rivers, increased risk of floods, and destruction of fertile farmlands, to name a few—have become apparent. Water renewal proponents, whose influence on water policy has grown considerably in the past few years, argue that it's cheaper to remove a dam than to try to ameliorate its adverse impact on the local economy and quality of life by refurbishing it.

Typical is the case of the Edwards Dam, which lay about forty miles upstream from the Atlantic Ocean on the Kennebec River in Maine. Everything about Edwards, a hydroelectric dam built in 1837, was tiny. It was about 102 feet long and three feet high, and it produced just one-tenth of one percent of the state's power supply. But this minuscule and even innocent-looking structure nonetheless did quite a bit of damage. Most notably, its undersize wall blocked access to the spawning grounds of hundreds of thousands of fish each year, including Atlantic salmon, American shad, river herring, sturgeon, bass, smelt, and eel. After assessing this problem in the mid-1990s, a federal energy agency ruled that the dam's negative impact on habitat and salmon fishing, a staple of the local economy, was much greater than its value as an electricity producer. Authorities mandated that ladders would have to be built to let the fish reach their spawning grounds. That sealed Edwards for extinction. Retrofitting the dam would have cost about $10 million; decommissioning it cost only about half that. In 1999 the Edwards Dam was destroyed. In a nice turn of phrase, then Interior Secretary Bruce Babbitt said that along with the dam, several myths were dismantled, including "that dams should last as long as the pyramids."

In other developed countries, dams are being destroyed for similar reasons. Notably, in France the Maisons-Rouge and Saint-Étienne du Vigan dams are being torn down, because both obstruct the spawning grounds of the Atlantic salmon on the Vienne River, the second most important tributary to the Loire.

The idea of restoring waterways, lakes, and rivers to a more natural state is a new one, only a few years old. But it's an extremely positive step. Rather than attempting to solve local water crises—such as a lack of water for irrigation and drinking, or a river polluted and diminished by a dam or diversion—by undertaking more management efforts, many regions are concluding that allowing water to make its own course, the way it did originally, might be more practical. Reaching this conclusion requires a starkly different attitude toward water and the environment than people have traditionally shown—it necessitates exchanging a sense of control, however false, for an acceptance of the unpredictability of nature. It's an implicit concession that in incident after incident water crises are the result of our own aggressive water management practices.

Tearing down a dam or restoring a water system that has been altered is expensive, however, and thus only well-off countries can typically afford to do so. For that reason, in emerging nations, which lack money and need to build their economies quickly (an activity that is still misguidedly linked with major hydroelectric and industrial water diversion projects), restoration isn't even a consideration. In fact, the opposite is taking place. Frequently backed by World Bank loans and advice, poorer nations are building bigger and bigger dams with no better environmental safeguards than the dams that are being destroyed in the United States and other developed countries.

The situation in the Everglades today should be warning enough for any country not to follow in our footsteps. But before we fault others for being slow to catch on, it's important to remember that the United States didn't even consider restoration to rectify the water management debacle in the Everglades—in fact, we continued to aggravate it—until it became clear that Florida was in danger of going dry.

That realization took place in the late 1990s, when Florida suffered one of the worst droughts in its history. With no precipitation year after year, except for enormous storms in the rainy season whose deluges were quickly ushered out to sea, conditions rapidly deteriorated. Lakes dried up, shallow municipal and private wells sucked air, and salt water crept into underground aquifers. As a result, orange groves and farmland were turning fallow, and thirsty alligators were prowling highways and back streets in suburban and rural neighborhoods for a sip of water. The possibility increased that some Florida cities would actually run out of drinking water. In desperation, the state considered measures that otherwise would have seemed unthinkable, such as pumping untreated waste water into underground aquifers, hoping that an experimental treatment process could clean and filter the water before it did permanent harm to water supplies, farmlands, and rivers. The most succinct description of the situation came from state climatologist James O'Brien: there were simply too many human beings placing too many demands on too little water.

The best evidence for how bad things had gotten was the sorry state of Lake Okeechobee, which (in addition to its role as an irrigation ditch for the rich farmland that borders it) serves as the backup reservoir for Florida's large east coast population. By mid-2001 Okeechobee was stagnant, shallow, and shrinking dangerously fast. It sat only about 9.5 feet above sea level, a huge drop from its twelve-foot level just a couple of years earlier, and even that was two feet below normal. Local water officials predicted that in the next few years, as things stood, Okeechobee would likely drop still further, to as low as seven feet. Should that happen, the lake's pumps would no longer be able to draw water to supply either the Everglades or local communities.

Magnifying the problem, Florida's population growth is increasing the drain on the lake. In 1975, four hundred years after Spanish settlers discovered Florida, the state's population reached 8 million; twenty-five years later, this figure had *doubled*. And another 8 million residents are projected by 2030.

Even with the blistering drought and growing population, however, Florida should have more than enough water to satisfy the needs of its many residents. The state is ringed with seven thousand lakes larger than ten acres, most of them near the huge Okeechobee; under natural conditions they could provide enough water for the millions of people living in Miami, Palm Beach, Boca Raton, and other eastern cities.

But there's no longer anything natural about Okeechobee, its satellite lakes, and the Everglades. Despite its beauty and uniqueness, the Everglades was always viewed as a nuisance in south Florida. It took up a lot of space that could have been used for other purposes, Floridians felt, like agriculture and new towns. And though the Everglades provided the residents with water, it also flooded inconveniently during rainy seasons, which discouraged even bigger increases in population. In 1947, a series of devastating floods from back-to-back hurricanes killed dozens in the area and submerged 90 percent of the land from Orlando to the Keys: the Everglades' fate was clinched. Determined to make sure a catastrophe like that would never happen again, the next year, with the help of the federal government, Florida began constructing a network of levees, canals, and pumping stations that would drain 750,000 acres south of Okeechobee, diverting its waters in a complex arc that at times fed sugar plantations and subdivisions and at other occasions spewed the water around the Everglades straight out to the sea.

The extent of Okeechobee's and the Everglades' subjugation is captured tangibly in West Palm Beach's South Florida Water Management District nerve center, a low-slung nondescript concrete government building standing halfway between the Atlantic and Okeechobee. It's so far away from either body of water that until you see the grainy black-and-white pictures of officials standing on the banks of rivers and lakes in the lobby, you would think this structure had nothing to do with anything wet. Tommy Strowd, a toothy, effervescent nine-year veteran of Florida's water reengineering effort, is the puppet master for the Everglades' watershed. In his first-floor control room, the size of a large living room, crowded with high-speed, 24/7 communication technology—computers, fax machines, weather service hotlines, and digital maps—Strowd oversees the flood gates and pump stations that control upward of two thousand miles of canals and levees, which shunt south Florida's water to towns and irrigation ditches that need it and send the overflow out to the bay. He manages an intricate system of microwave telemetry that reads water activity everywhere in the region minute to minute by canvassing outdoor gauges. After the computers process the data, they make suggestions for which pipes and gates should be opened and which closed, and how quickly. Strowd then handles these tasks, moving the water to where it's supposed to be—at least according to the computers—by remote control.

One of the most striking things about Strowd is how oblivious he is to the real meaning of the system he has at his fingertips. In effect, merely by reading a computer screen and pressing a few buttons, he's physically changing the course of nature every day. It's a remarkable, if somewhat disturbing, demonstration of what people are able to do technologically, but it's also a distressing illustration of how much more comfortable and intellectually capable we are with the mechanical than with the environmental, and of how easy it is to distance ourselves from the obvious consequences of our actions.

"We created a totally controlled system," Strowd said almost casually, chobee region in 1960 and purchased four thousand acres of land for $160 per acre. In just forty years, their companies, Florida Crystals and Flo-Sun, now controlled by their children, have been transformed into a sprawling conglomerate with three thousand employees and 180,000 acres of land that produces more than 10 percent of the nation's sugar. At least $65 million a year in the Fanjul family's earnings comes from government price supports.

The Fanjuls, who are worth upward of $1 billion, have returned the generosity by making extensive contributions to both political parties. During the 1992, 1996, and 2000 presidential election campaigns, the Fanjul family gave more than $575,500 to campaigns for federal offices, and their sugar companies donated at least $843,000 directly to Republican and Democratic national committees, according to Federal Election Commission records. These contributions have apparently helped the Fanjuls—who are extremely well-connected, so well, in fact, that in February 1996 President Clinton even interrupted a meeting with Monica Lewinsky to speak with Alfonso Fanjul—lobby successfully against changing price support programs and for protecting the sugar farms, even as the Everglades is restored and more water is funneled to east coast cities to relieve their shortage.

In fact, the tension between the two factions of South Florida—Okeechobee farmers and the east coast metropolitan areas—has always been the stumbling block to any effort to restore the Everglades, a movement that began in the early 1990s, when the environmental harm that had been done to the ecosystem was becoming more and more clear. In the late 1940s, rerouting Okeechobee and the Everglades to stop flooding and allow for development created the vast sugar farms and enabled the East Coast cities to stretch out, but at the same time cut off south Florida towns from a substantial part of its water supply. The project's elaborate plumbing system requires Okeechobee to squirt some 1.7 billion gallons of south Florida water into Florida Bay every day, so that it's kept away from surrounding communities. That's enough water to support more than 15 million people. This water diversion wasn't particularly noticeable as long as the region was still getting by on its primary source, the underground Biscayne aquifer. Consequently, at that time, arguments about the environmental damage inflicted on the Everglades didn't resonate with either region.

But when the drought occurred and the Biscayne aquifer began to dry up, Lake Okeechobee was supposed to be a dependable backup. Instead, it continued to conscientiously send its excess water into the ocean, while itself withering from the lack of rain. That's when it became impossible to ignore that the attempt to control the Everglades—in effect, to shunt it out of people's way—had backfired. Certainly, doing so had created wealth on the sugar farms and in billion-dollar cities, but that only increased the need for more and more freshwater for the region—and the Everglades project was specifically designed to get rid of freshwater. As a result, Florida was drying up. It was a paradox, to be sure, one so vivid that finally everyone in Florida began to "see things globally," as water engineer Tommy Strowd would put it. The Everglades had to be restored, or south Florida would become an economic and social ghost town.

The plan to restore the Everglades is contained in a four-thousand-page report by the federal government that, as detailed as it is, doesn't give a clear picture of what the Everglades will look like in forty years or so, when the project is completed. The goal is to reimplement the natural water cycle of flow, flood, and drought that existed until the late 1940s, when the Everglades was harnessed. And since a restoration this ambitious has never been tried before, no one is sure whether accomplishing it is even possible.

The key element of the effort involves capturing most of the 1.7 billion gallons of water that is now pumped daily from Okeechobee into the ocean and storing it in reservoirs. From there, it will be distributed to the Everglades, as well as to sugar farms and to east coast communities. Dikes and barriers at the eastern edge of the Everglades will be torn down, giving the ecosystem room to expand even if it encroaches a bit on south Florida metropolitan areas.

As for the sugar farmers, initially it appeared that they could lose as many as 200,000 acres, or nearly one-third of their land, to reservoirs and Everglades reclamation. But after mounting a formidable lobbying campaign in Washington, they succeeded in getting lawmakers and the Clinton administration in late 2000 to approve a program with a decreased estimate: 60,000 acres, a mere 8 percent or so, will be affected.

The Everglades restoration project didn't come about because of environmental concerns over the destruction of a delicate and absolutely essential ecosystem. It occurred more because of financial fears—that Florida could not remain economically viable and attract more business and development if it were running out of water—and the sudden eye-opening realization that turning on the tap in Florida and getting nothing out of it was a real possibility. Still, the fact that restoration is occurring at all carries weight. Just ten years ago, no one would have seriously predicted that the Everglades might be returned to its natural state, and few would have argued that it should ever be undertaken. Now that it's taking place, or at least being attempted—after conceding that we made a huge blunder in thinking we could manage water so aggressively—other water restoration projects no longer seem as outlandish as they once did.

This is critical because even as the Everglades is being restored, carelessly planned and potentially devastating water management efforts like South America's Paraguay-Paraná Hidrovia project are under consideration. The 42 million acres of the Pantanal, located in Brazil, Bolivia, Paraguay, Argentina, and Uruguay, constitute the world's largest inland wetland, seventeen times larger than the Everglades. Officials of these nations have put together a plan to create thousands of miles of barge canals in the Pantanal, in an effort to boost industrial production and trade. The project would require dredging, deepening, straightening, widening, diking, and damming this interconnected system of rivers and marshland.

The Pantanal is currently undeveloped, but it is already being polluted by millions of gallons of untreated waste that are discharged directly into the rivers that feed it. Turning it into a barge canal would not only ruin its fragile ecosystem, destroying aquatic life and spreading disease through the region, but would also severely diminish local water supplies.

Another similarly irresponsible project is under way in Lesotho, a tiny and poor nation of 2 million surrounded entirely by South Africa. Lesotho is spending hundreds of millions of dollars to divert a significant portion of its only freshwater supply, a series of rivers in the Maloti Mountains, to

South Africa's industrial Gauteng Province. Lesotho wants to create a consistent revenue stream by selling this water to its neighbor and also hopes that jobs will be created in South Africa for which Lesotho residents might apply. But there are serious consequences to damming the water in the mountains and building pipelines to divert it elsewhere—especially since this is a vital water supply for Lesotho. Reduced flow rates and less frequent floods in the mountains' rivers will almost certainly pollute the waters and turn a once vibrant and pure water supply into a stagnant, useless pool.

After all the water management failures and their dreadful repercussions that we've experienced, it is bewildering that projects like the Paraguay-Paraná Hidrovia and Lesotho could still be under serious consideration. I've struggled to understand why, and the reason, I've come to believe, isn't a particularly creative one, nor even one that has been proven to be correct: many countries in the world still suppose that a major water management effort is a quick economic fix that can instantly elevate a nation into the upper tier of the global economy.

What else could explain projects like China's Three Gorges Dam, the largest and most powerful dam ever attempted? Under construction across the Yangtze River and slated to be finished in 2003, the Three Gorges will be more than six hundred feet tall and create a nearly seven-hundred-mile-long reservoir in an imposingly beautiful stretch of canyons that has inspired the work of Chinese poets and artists for centuries. China is counting on the 18,000 megawatts of hydroelectricity produced by the Three Gorges—50 percent more than Brazil's Itaipu Dam, currently the world's largest hydroelectric dam—to be the backbone for countless industrial projects in the region. In addition, China hopes that after it is dammed, the Yangtze will become a thriving barge canal for extensive world trade.

Critics, among them a group of fifty-three leading Chinese engineers and academics, publicly doubt that any of these benefits will be realized for a long time, because the Chinese economy is simply too small and lacks the capacity to meet these forecasts. At the same time, the potential environmental and social damage of Three Gorges is tremendous. More than one million people will be forced to evacuate their homes as a hundred towns are submerged; 35,000 acres of agricultural land will be drowned; outbreaks of water-borne diseases like malaria and the destruction of fish habitats will occur downstream as the dam's reservoir lowers the Yangtze's depth; and hundreds of archaeological treasures will be flooded, such as a seventeen-hundred-year-old temple and the tombs of the Ba people, who lived more than two thousand years ago. The Yangtze is a dangerous river to tamper with; more than 350 million people live in its watershed. But Chinese officials, ignoring all of this and publicly scoffing at the dam's critics, some of whom designed its original plans, have refused to rethink the project and are even threatening to speed up its construction.

The more I think about the environmental blind spots and the recklessness exemplified by Three Gorges, the Paraguay-Paraná Hidrovia, Lesotho, and so many other projects, and the more I consider the anxious effort in south Florida to reverse the destruction of an ecosystem before the region goes thirsty, the more I wonder what water calamities await China, Bolivia, Paraguay, and nations involved in similar misguided efforts. I have little optimism that in fifty years these countries will be able to afford the tens of billions of dollars they will need to restore the precious water resources they are so intent on devastating now.

This adds yet another significant wrinkle to the water crisis. If we look at today as a snapshot in time—that is, as if the problem of water scarcity were to hold at the current unacceptably high level long enough for us to do something about it—it would take a remarkable amount of technological skill, creativity, and financial resources to accomplish the goal of supplying enough water to enough people so that the crisis would be viewed merely as controllable. But with every new misguided water project we attempt, we're adding the burden of billions of additional dollars and more and more difficult engineering feats to the solution, while dangerously telescoping the amount of time we have to implement one. This could be a deadly combination for the planet, and it highlights the need to attempt what for us has been impossible until now: to consider water projects cautiously, instead of with abandon—meticulously analyzing them for environmental as well as economic consequences—before we make any more mistakes that we will be forced under pressure to try to undo.

7

There Are No Winners

We desperately need to recognize that we are the guests not the masters of nature.

MIKHAIL GORBACHEV, PRESIDENT, GREEN CROSS INTERNATIONAL

IN APRIL 2000, I received an invitation to attend a speech by then Secretary of State Madeleine Albright to commemorate the thirtieth anniversary of Earth Day. Dozens of similar events took place that spring, but this one was intriguing because of the odd location Albright chose for the address: the National Defense University at Washington's 210-year-old Fort McNair, a college that specializes in training antiterrorist operatives, CIA analysts, and spies—in other words, a school more associated with James Bond than with John Muir.

Just minutes into Albright's speech, it became clear why she had selected NDU and, in fact, why she—a political scientist not known for environmental positions—felt the need to make an Earth Day pronouncement in the first place. Albright, like many other international security experts, was alarmed about the water conflicts that were recently breaking out on virtually every continent and the possibility that they could explode into hostilities that threatened some of the world's most vital but unstable regions. With NDU's reputation as a think tank that analyzes strategic responses to international differences, what better backdrop to make the point that fighting over water must be taken as seriously as even the bloodiest civil war?

In her address, Albright returned to this theme repeatedly. At one moment early on, her oratorical delivery suddenly became conversational, and she leaned forward on the podium, pulling the audience closer to her, and said, "Water will be the main focus of my remarks today. I have chosen this topic because, although water is often thought of in very local terms, it is certain to be among the principal global challenges of the twenty-first century."

This, it struck me immediately, was an extremely important statement. It was the first time I had heard a minister of state anywhere in the world publicly suggest that water is, or must be, a strategic imperative and a priority for the international security community. In doing so, Albright also tacitly acknowledged that water conflicts have become bedeviling, even to people who spend much of their time negotiating complex treaties and keeping difficult alliances from unraveling. "As a diplomat, I have seen firsthand the tensions that competition for water can generate," Albright said. "Further disagreements over access and use are likely to erupt."

Albright couldn't have been more correct. In countless flashpoints around the world, more than ever before, people are now threatening each other over who gets how much, and for what purpose, of the limited amount of available water. Many of these disputes are centered in the more than three hundred river basins and aquifers that are shared by two or more countries: places like the Mekong River, which rises in China and courses through Myanmar, Thailand, Laos, Cambodia, and Vietnam—nearly a quarter of a billion people live in its watershed. So far, amid harsh words, there's been no agreement among the Mekong nations on allocating the use of the river's waters. Meanwhile, as the countries jostle to wrest the biggest share of the shrinking river, dozens of dams have been built—rapidly turning the river into a polluted and tepid waterway. In Africa, all fifty-four water bodies that pass through the borders of more than one nation—including the world's longest river, the Nile—are the center of similarly tense disputes. And the Jordan River may be the most stressed watershed of all, geopolitically separating Israel, which effectively controls the river, from friendly nations such as Jordan and from enemies such as Palestine.

Many of the most troubling conflicts are not simply quarrels over access to water but thoughtless episodes that also carelessly damage the environment. The overdamming of the Mekong is one example, but two 1991 Gulf War incidents are even more eye-opening. One took place as U.S. soldiers were chasing Iraqi troops out of Kuwait. Though Saddam Hussein's forces were retreating and defeated, they took one last swipe at disabling Kuwait by destroying its desalination plant, which removes salt from Persian Gulf water and turns it into freshwater that can be used for drinking, bathing, industrial purification, and the like. Kuwait has no lakes and rivers and depends on desalination as well as water shipments for virtually its entire water supply. By demolishing Kuwait's desalination capabilities, Iraq hoped to strangle its neighbor financially, forcing it to search for new and more expensive sources of water, and at the same time to punish its residents, who would have had to survive on much less water.

For a time Iraq succeeded. It took Kuwait more than two years to rebuild the desalination plant. During that period, the country had to spend billions of dollars to purchase a minimal amount of water from friendly countries, mostly other water-scarce Arab nations, that were willing to share a bit of their supply. Meanwhile Kuwaitis bickered and blamed their government, even to the point of street rioting, over the much more limited amount of water that was available. Seeing how effective this strategy could be, two years later Saddam poisoned and drained the water supplies of Shiite Muslims in the south of Iraq to suppress opposition to his regime.

In another Gulf War incident, the United States matched Saddam's tactics. To incapacitate Iraq—and demoralize Iraqis—the United States and its allies rained dozens of targeted bombs on the country's water supply and sanitation systems. The wreckage was widespread, and much of Iraq's water infrastructure was leveled. Even today it hasn't been completely rebuilt: there are many pockets in the country where clean water is yet unavailable.

These little-known Gulf War episodes are distressing for reasons that go beyond the mindlessness of war. The attacks represent a level of irresponsibility that the people involved—mainly Iraqi and U.S. officials—never seemed to consider: at their hands, over a period of about seven days, two precious water systems in the Middle East were destroyed, heedlessly and without remorse, in a world that is running out of water and can't afford to lose any more.

Unfortunately, the idea of recklessly endangering a limited supply of water to punish an enemy is not original to our time. As long ago as 1700 B.C., Abish, grandson of the great Babylonian leader Hammurabi, tried to trap and flood retreating Mesopotamian rivals by damming the Tigris River. He was too late. By the time the dam was built, permanently altering the Tigris's flow, the Mesopotamians had already left the area, only to return soon afterward with reinforcements that would eventually overrun Abish's armies and mark the decline of the Babylonian civilization.

Placed in their fullest context, all of these actions and their ramifications yield an inevitable conclusion. That is, while much of the mismanagement of water throughout history has involved over-the-top construction efforts and questionable water diversion schemes—usually in the name of profit, economic development, and sprawl—an equally damaging parallel thread of activity links the events in the Gulf War and the Mekong Delta to the destructive damming of the Tigris almost four thousand years earlier: without regard to the environmental implications, water has been repeatedly thrown like a grenade into the crossfire of conflicts big and small, and cavalierly used as a disposable weapon by which one group gains physical or economic power over another.

◖ Albright's Earth Day speech, which lasted a half-hour, was passionate and pragmatic, with touches of poetry that moved the few hundred of us in the audience—especially in her closing, when she said: "From history's dawn to this morning's, wells and streams, rivers and lakes, have meant life. Every great civilization has grown up around water. From the Ganges to the Mississippi, the Amazon to the Zaire, the history of rivers is the history of us. And there is no more unifying or naturally democratic force.

"Creeks formed in the highlands of every continent gather strength in their journeys to the sea. And as they flow, channeled by swerve of shore and bend of bay, they cleanse, nourish and refresh all people—in metropolis and village, from the millionaire to the child who knows no other cup but the human hand.

"Today, this irreplaceable resource is in irrefutable danger. For too many, the liquid we cannot live without bears within it the cause of illness, even death. It doesn't have to be."

It was a masterly stroke for Albright not only to link protecting water supplies with international security but also to end her Earth Day address with an evocative rendering of the world's water crisis—especially to portray the increasingly limited amount of water and the pollution and decay that people have instigated. Water wars—whether pitting nation against nation, or neighbor against neighbor—are both a symptom and cause of the larger water crisis. Think of it this way: When people vie for a diminishing supply of water, the fear that there won't be enough to satisfy their needs intensifies. As a result, scarcity itself makes water an intimidating weapon in the hands of those who have it or control it. But it's a vicious circle. The environmental damage—the degrading of water systems, the poisoning of rivers and aquifers, and the disruption of the natural course of water—that characterizes water conflicts magnifies water shortages. As that occurs, just the threat of depriving a neighbor or nation of water becomes even more menacing.

A perfect example is the widening rift between Singapore and Malaysia, Southeast Asian rivals that were members of the same federation for a few years until Singapore seceded in 1965. Singapore is densely populated, with about 5,500 people per square kilometer. As an island nation, it has a handful of local streams and reservoirs but far from enough freshwater for its 3.5 million residents. By contrast, Malaysia, directly across the narrow Johor Strait from Singapore, has fewer than 70 people per square kilometer and an ample supply of water from its two large rivers, the Rajang and the Kinabatangan, and several reservoirs.

Under an agreement reached decades ago when relations between the two countries were more amicable, Malaysia supplies Singapore with 350 million gallons of water each day—well over half of Singapore's water supply—through an underground pipeline. This arrangement, which will expire in stages through 2010, has given Malaysia an enormous advantage over Singapore, one that Malaysia is using more and more often. As the two nations have bickered over, for instance, Singaporean investments frozen by the Malaysian government during a recent economic crisis or religious differences between Singapore's Hindus and Malaysia's Muslims, Malaysia has warned of cutting off Singapore's water supply if it doesn't tone down the rhetoric and capitulate to Malaysian positions. In a particularly tense moment, a high-ranking Malaysian minister even threatened to seize two of Singapore's water purification plants, which are located on Malaysian soil.

Just the possibility that Malaysia might sabotage Singapore's water supply cemented Malaysian influence over Singapore in diplomatic and economic matters. Singapore's population is forecast to double in the next twenty-five years to about 7 million; its very survival would be jeopardized if Malaysia gained control of its water plants, an act that would, in effect, put the spigot that regulates the flow of water between the two countries in Malaysian hands. These days, that prospect can be as frightening for a nation as that of having a nuclear bomb pointed at it.

A potentially nastier water conflict, one that is getting achingly close to bloodshed, is simmering in southern Africa between Namibia and Botswana. These two countries share the Okavango River, which rises in some of the world's most densely wooded savannahs in Angola and eleven hundred miles later empties into Botswana's Okavango Delta, a swamplike ecosystem with a diversity of plant and animal life unrivaled in Africa. During its course, the Okavango passes through a tiny sliver of Namibia, known as the Caprivi Strip, before entering Botswana. How much water Namibia and Botswana are allowed to take from the Okavango has long been an ongoing point of contention between the two nations. But neither has ever threatened to breach their tenuous supply agreements until recently.

Because the Okavango traverses its territory first, Namibia holds the upper hand. In the late 1990s, the nation turned this asset against its neighbor. While engaged in a separate dispute with Botswana over the fate of two islands—both of which offer another possible supply of freshwater—Namibia drew up plans to construct a 155-mile pipeline to divert from the Okavango as much as 100 million cubic meters of water per year, more than tripling the amount of water it now gets from the river. Under this proposal, Namibia would ship the water southwest to Windhoek, its capital city, where water shortages during drought years are severe and where in the countryside recently the corpses of sixty thousand head of cattle that died of thirst were left to rot. But if Namibia intercepted this much Okavango water, it could cripple Botswana's water supply. It would also irreparably ruin the ecosystem in the Okavango Delta, which besides being an irreplaceable animal and plant habitat is also a hugely popular tourist attraction in Botswana.

Botswana viewed Namibia's water plan as an act of aggression, and many people in the region felt Namibia was threatening to take more of the Okavango as a way to convince Botswana to give up its rights to the disputed islands. Instead of capitulating, though, Botswana escalated the conflict by wildly increasing its military spending. It began construction—with the help of the U.S. Central Intelligence Agency, according to observers in the region—of a secret military air base in its southern zone that can handle the most sophisticated modern bombers. And it has purchased more than a dozen U.S.-built fighter-bombers from Canada as well as tanks, troop carriers and massive amounts of medium-range artillery.

So far this buildup hasn't scared off Namibia, which is standing its ground and says it still intends to divert the Okavango. Fearing that things could turn ugly, South Africa, which borders both countries and at one time counted Namibia as one of its territories, has tried to negotiate a peace between the nations—so far, to no avail. Without an agreement soon, though, the stability of the region will be seriously undermined, says Peter Mokaba, South Africa's deputy minister for environmental affairs and tourism. "There can be major economic, political and even military consequences for the area unless this gets resolved," he told me.

As difficult as these complex water conflicts are to resolve, the threats at least have a sense of premeditated geopolitical strategy and logic. That's decidedly not true, however, of dozens of other more random and more perversely violent acts involving water, most of them revealing little else but the darkest side of human nature. There's the incident in 1999 when the Indonesian-backed militia opposing East Timor's independence killed thousands of guerrilla troops on the island and then purposefully threw the bodies in local water wells to pollute the aquifers. That event was not isolated. It's a tactic the Serbs used around the same time in Kosovo, after a particularly gruesome rampage that murdered tens of thousands of Kosovar Albanians. The corpses were dumped into the province's wells, and within hours blood was pouring out of Kosovo's water faucets. And in a terrorist action in 1998, a guerrilla commander in Tajikistan planted a bomb at a dam on the Kairakkhum Channel, threatening to drown tens of thousands of central Asians if his demands weren't met.

Water conflicts may affect the future of the planet, but they're not global disputes and generally they can't be resolved by global intervention. In fact, what makes them so challenging is precisely that they are local skirmishes—hundreds and hundreds of them in emerging as well as developed countries, built on any number of competing layers of self-interest unique to the region that they're being fought in.

By contrast, in the overall water crisis itself, any number of international initiatives can be undertaken to alleviate water scarcity and pollution and to undo the negative consequences of unnecessary dams and misguided diversion efforts. Worldwide funding policies can take into account environmental and social impacts; innovative technologies can be developed to deliver clean water to places that need it; market-driven water pricing can subsidize supplies in poorer countries with money paid by consumers in wealthier nations; and the world community can make an international commitment to the idea that everyone has the right to freshwater. But by their very nature, water conflicts have much narrower boundaries than these comprehensive public and private efforts. They tend to pit one against one—even when *one* means "many"—face to face, my need versus your need.

Former Soviet Union president Mikhail Gorbachev, who is building a stellar reputation as a mediator in water conflicts, says he learned that water disputes are provincial during a series of strange incidents in the

1970s, when he was a member of the Politburo under Leonid Brezhnev, overseeing agricultural matters. Four large Soviet republics in Central Asia relied on two rivers, the Syr Dar'ya and Amu Dar'ya, to irrigate their extensive rice fields and to supply their fast-growing, extremely poor urban populations. One of Gorbachev's jobs was to allocate water quotas to these republics and settle any disagreements that arose over the amounts he allotted. No decision he made, Gorbachev says, could satisfy everyone, and more often than not the peasants would end up physically fighting with each other to get more water.

"What often happened is, let's say, during the daytime an agreement is reached and quotas are allocated and the machines are adjusted for this allocation," Gorbachev says. "But then overnight the peasants changed it and readjusted the allocations made by the machines. The strongest peasants, in the end—the ones with the most people and weapons on their side—got the most water. No outsider could imagine what was going on here or what to do about it. You had to be there, thinking the way the peasants thought, to handle this and fix it. Sometimes we had to give more to one group and promise another one more the next time. Sometimes we succeeded in making everyone happy, sometimes we didn't."

Inspired by the difficulty of these negotiations as well as a sense that resolving water disputes locally is critical to the planet's survival, in 1993, soon after ceding power during the fall of the Soviet Union, Gorbachev founded Green Cross International, an organization whose goal is to prevent and mediate conflicts in what he calls water-stressed regions. The technique that Green Cross most favors is to continually maintain negotiations between the people involved in a dispute—in other words, make sure that there are continuous meetings between the parties—even if heels are dug in and there's very little change in position from one session to the next. Although this approach is tedious and can take years to produce limited returns, its benefit is that adversaries are always face to face, in the same room, and less apt to take rash action. "And when we have them in front of us, we remind them over and over that we are the guests not the masters of nature," Gorbachev says, using one of his oft-repeated expressions. "We educate them that they don't have the right to use water as a weapon against each other, because water is not something that can be possessed."

Green Cross's biggest success so far has been in South America's Pilcomayo River, a sixteen-hundred-mile-long tributary of the Paraguay River that rises in Bolivia's Andes Mountains and flows southeast to form the border between Argentina and Paraguay. More than 90 percent of the Pilcomayo is in Bolivia, near some of the country's most plentiful mines. That geographical imbalance has led to disconcerting environmental consequences. In Bolivia, runoff from gold, silver, tin, lead, and zinc operations as well as untreated human waste from mining community residents pollute the Pilcomayo, turning it into a muddy and slothful stream long before it reaches Argentina and Paraguay. But in Argentina and Paraguay, native populations like the Guaraní and Mataco tribes drink the filthy water straight from the river, unpurified, and use it for irrigation. As well, a staple in their diet is the river's savalo fish.

For years, the Indians tried to draw the Argentine and Paraguayan governments' attention to the squalid conditions of the Pilcomayo, but they were mostly ignored. Then in 1996 an accident occurred that nearly led to violence among the three nations. A dike in a Bolivian mine suddenly burst, sending 300 cubic tons of tailings into a tributary of the Pilcomayo. By the time the Pilcomayo reached Argentina and Paraguay, the damage couldn't be ignored: the river was covered with metal sediment, including high concentrations of poisonous lead and arsenic that were well beyond international water quality guidelines for human consumption.

Fearful that water supplies throughout their territories—and not just the water drawn by the Indians—would be affected, Argentina and Paraguay threatened to strike back at Bolivia if it didn't do something immediately to stop the pollution of the river. Bolivia, hoping to avert war with two countries at once, instituted an emergency plan to repair the mine's dike, dredge polluted sediments from the Pilcomayo, and monitor water quality.

That worked, to a point, and tension was eased. But Bolivia followed through on these steps only half-heartedly—for instance, it stopped measuring for toxins in the river within months of promising to do so, and the Pilcomayo remained extremely polluted. Worse yet, through diplomatic channels, Bolivia, which was cash-strapped and couldn't afford to fully clean the Pilcomayo or keep the poisons out, took an aggressive gamble by telling Argentina and Paraguay that the river was essentially owned by Bolivia and that it would decide its fate.

In late 1997, with tempers pitched and the possibility of border skirmishes escalating, Gorbachev asked to try to mediate the standoff. The three nations agreed, and after a series of lengthy negotiating sessions, some of them lasting weeks, Bolivia, Argentina, and Paraguay signed a Green Cross–brokered plan that would eventually ensure that the Pilcomayo will be pollution free, though not immediately.

Under the program, the first few years, ending in about 2001, would be used to identify with sophisticated equipment which wastes, canals, dams, and discharges on the Pilcomayo were actually toxic and to what extent. No investigation of this sort had ever been done before. After that's finished, the next step will be to draw up a schedule that requires Bolivia, with the help of the other nations, to initially reduce and eventually eliminate the river's poisons based on what was learned during the fact-finding phase. The Green Cross compact explicitly mandates specific funding responsibilities for the program. Money is slated to be provided by the three South American countries in equal amounts as well as the World Bank and numerous environmental organizations. Experts from the nations are expected to work for free or nearly so. Perhaps most noteworthy, Bolivia, Paraguay, and Argentina acceded to Green Cross's demand to drop any threats of violence over the Pilcomayo during the life of the agreement.

The impressive simplicity of the Green Cross Pilcomayo accord is also its genius. In effect, what Gorbachev and his colleagues achieved was a cease-fire without anybody agreeing to do anything more than just collect information and keep telling each other what they found out. And as importantly, he convinced the three South American nations to publicly accept that they share the same goal—namely, to clean up the Pilcomayo.

"It's an example of how to deal with water conflicts through knowledge," Oscar Natale, head of Argentina's toxic substances program at its National Institute of Water, told me. Natale says he's still in awe of how easy it was to get these three angry countries to drop their guard and, once they were face to face, agree to much less than they initially demanded. "We realized it was to our own benefit to come out with something positive in these meetings; the alternative suddenly became unacceptable. Now all we're doing is finding out things that we didn't know before, like what is toxic in the river and how much of these toxics there is. So we don't need to act out of emotion or misinformation anymore. When we're finished gathering the truth, we can do what we have to do next, which is clean up the Pilcomayo. The best part, for the first time in as long as I can remember, we are optimistic that we won't fight over the river and that the river will soon not be polluted."

Water conflicts in developed nations may not have the aura of impending violence that exists in emerging and less established countries—although it was only about eighty years ago that Owens Valley residents blew up pipelines to stop Los Angeles from seizing their water—but these disputes are just as bitter and just as common. In wealthy and poor countries alike, conflicts almost always erupt from the fear of not having enough clean water, and very few regions in the world are insulated from this fear. If mounting demands from population growth, development, and industrial expansion outstrip supply, no amount of money can provide adequate water. Consequently, in the United States, for instance, states—and even people who live next door to each other within states—are increasingly bickering over shared water sources, usually to get a little more before it is all gone.

In most cases, these conflicts are fought in a hyperactive political and economic atmosphere, where no one seriously discusses putting brakes on population growth and corporate development—two factors that increase the need for more water. Only lip service is given to limiting sprawl, mandating conservation, or managing water supplies more intelligently to eliminate the cause of the dispute. As a result, many fights over water in developed countries are never really settled—they're just buried for a time under a short-term fix and will inevitably emerge again as the amount of available water lessens and the need for it heightens. Ironically, water sharing agreements among emerging countries, which usually include clauses to clean up severely depleted water supplies precisely because there is so little in the first place, are often much more environmentally sensitive than those in developed countries, where protecting the quality of the water supply is frequently a secondary consideration.

This has been true in the dozens of water conflicts that continue to pepper every region of the United States. Virginia and Maryland, for instance, have been fighting over the Potomac River in a case that traces back to 1632, when King Charles I of England granted Maryland ownership of the Potomac, from one end to the other. Despite that colonial claim, Maryland has let Virginia tap Potomac water, but only at its own shoreline. Recently, Virginia asked to extend its intake pipes about seven hundred feet, into the middle of the river. The shallow waters near the Virginia banks have become too dirty for use, even after purification; more usable water is available in the deeper portions of the river. But Virginia only has itself to blame for its clean water deficit, says Maryland state senator Christopher Van Hollen: "There's just been a lot of churning up of the land to make room for lots of new development and roads and highways. They've mucked up their part of the river, and now they want to take more of ours."

Despite these sentiments, this dispute is more about money than about clean water. Virginia says that by siphoning water from the middle of the river, it will save about $1 million annually in filtration and treatment costs. Maryland officials, meanwhile, are convinced that Virginia will pass along these savings to consumers and businesses. They worry that cheaper water will accelerate the industrial and residential development of Virginia's Fairfax County and stymie growth in Maryland.

Under pressure from federal officials, who didn't like the specter of states fighting so near to Washington, D.C., Maryland finally agreed to let Virginia extend its pipeline into the Potomac, but it refused a request to allow a bridge across the river that would open access to Fairfax County's high-tech corridor. That part of the disagreement may have to be decided by the U.S. Supreme Court.

In Texas water is literally becoming scarcer by the day, a fact symbolized by the sorry shape of the once vital Rio Grande, so depleted by over-use and diversion for irrigation, industrial purposes, and drinking water that it now stops fifty feet short of the Gulf of Mexico. Residents of the Panhandle are now cutting deals to sell the water that lies under their land to parched cities such as El Paso, San Antonio, Dallas, and Fort Worth. Billions of dollars are at stake in these transactions, which would siphon water from a mostly untapped section of the Ogallala Aquifer, a huge underground reservoir that stretches through eight states of the Great Plains as far north as South Dakota. In certain areas of the country, the Ogallala is being drawn down a hundred times faster than its ability to replace the water, leaving farmlands and towns drier and drier.

The prospect of diverting Ogallala water to distant Texan cities has set off a bitter argument between the state's ranchers and the would-be water entrepreneurs in northwest Texas near the Oklahoma border, one of whom is T. Boone Pickens, the oil tycoon who made huge amounts of money in energy company takeover bids during the 1980s. According to projections, if all the potential water sales occur, the Ogallala could dry up in as few as fifty years, especially if recent drought conditions persist. That would turn the region into a Dust Bowl, eliminating the ranches in the northwest as well as a good portion of Texas's overall water supply.

"You're going to devastate a large part of the state," Tom Beard, a rancher in Roberts County, told me. "Water isn't oil. We can't pump it up, use it up, and then try to find it someplace else."

Pickens, who wants to build a pipeline that can pump as much as 60 billion gallons a year, disagrees with Beard but says his analogy is an apt one. "Oil is the lifeblood of the world; people need it, so you need to keep providing it," Pickens says. "Water is the lifeblood of West Texas. The cities have to get it from somewhere."

The legislature and state water agencies, meanwhile, are at a loss as to which side in this dispute to favor. "If we make a mistake on this one, we affect generations to come," says C. E. Williams, manager of the conservation district in Roberts County. "That's what makes me lie awake at night."

More than anything, the sheer selfishness that seems to pervade water conflicts in the United States dismays me. In places where water supplies are already at a premium and where pollution is already a concern, people have simply tried to squeeze more water out of a drying pool and get their siphon in the ground before their neighbors do. Yet as perplexing as these disputes are, almost everybody I spoke to told me that none of them matched the obstinate attitudes and dug-in enmity in the battles that have ruined the rivers of the Northwest and made the Pacific salmon virtually extinct.

The Skagit River is as typical a roving Washington state river as there is. It meanders through coves and sloughs, dallying to add miles and time to its journey that begins in the snowy Cascade Mountains of the north, as if it were doing everything it could to avoid being swallowed by the salt water of the Pacific Ocean. With its relaxed course and the thick dark stands of fir on its banks, the Skagit is peaceful, a vessel of sunlight, perfect for sitting back on a boat and listening to the waters' lap. It is a fisherman's dream, although without many fish.

Salmon are tethered to the Skagit. Nearly three thousand streams feed the river during its more-than a-hundred-mile course, and the Skagit accounts for 20 percent of the water that empties into Puget Sound, Washington's sprawling ocean inlet. All six species of Pacific salmon spawn and then immediately die in the gravel bottoms of the Skagit. After a short period, their offspring embark for the ocean through the sound. As many as five years later, they come back upriver to deposit their eggs and revive the cycle.

Just a few decades ago, annual salmon runs would turn the Skagit orange, swarming with fish that looked to be flowing through veins as they hurriedly returned from the ocean. Now every Skagit salmon species is either on the endangered list or has been designated as threatened with extinction. During a typical spring run as many as ten thousand Chinook—also known as king salmon because of their size; a Chinook can grow to about four feet long and weigh as much as fifty pounds—used to return to the Skagit; now, fewer than a thousand do.

It's the same in all the rivers of the Northwest, especially the biggest ones like the Snake River, which rises in Wyoming and joins the Columbia River as a tributary in Washington. In the early 1800s, when the explorers

Lewis and Clark described the Snake as "crouded [*sic*] with salmon," about 2 million adult fish made the trip back to their spawning grounds. But more recently, only about eight thousand spring and summer Chinook, three hundred fall Chinook, and *two* sockeye returned to the river from the ocean.

The main culprits are the dozens of dams, most of them built in the last fifty years—almost all to produce hydroelectricity—that obstruct virtually every river in the region. There are five dams alone on the Skagit. These obstacles make it virtually impossible for salmon to go either upstream or downstream. Incredibly, federal and local authorities have spent as much as $3 billion on a series of so far fruitless and often preposterous schemes to help salmon get around the dams. There are now intricate fish ladders that hoist the salmon one level at a time, higher and higher, until they reach the peak of the dam; then the fish are tossed overboard onto the other side. And when that's unworkable because the dam is too high, even stranger techniques will move the salmon from one place to another. I observed one of them on the Snake River.

At a dam with a placid reservoir behind it and a miniature fall in front of it as water flushed over the concrete barrier, I watched with bemusement as a state fisheries biologist plucked one six-inch salmon after another from a trough into which they had been maneuvered just as they reached the obstruction in the water. He measured each fish, typed a couple of comments into a laptop computer, and placed the fish in a second bin that would be loaded onto a tank truck. The baby salmon were then hauled 250 miles over land, past a half-dozen dams, at which point they were returned to the river to complete their journey to the ocean. In its brochures, the federal government describes measures like these as "custom-made" to make the salmon's trip around the dams "as safe and comfortable as possible."

Dams aren't the only problem for the Pacific salmon. The widespread development of Puget Sound and other inlets and estuaries—mainly the expansion of farms and the building of pricey housing, hotels, and restaurants on prime property with an ocean view—has also been harmful. When younger salmon arrive at the ocean, they retreat to deep slow pools for protection while they mature, before facing the more dangerous sea. But the construction boom has engulfed those pools. Overfishing by aquaculture companies and local Indian tribes have also taken their toll. A Skagit-born salmon can be caught when it's in the ocean—as far away as Alaska—or in the river, on its return trip to its spawning ground. Considering this wide geographical range, in years past as many as 75 percent of adult Skagit salmon were harvested before making it back to the rivers to produce offspring.

This grim situation has abiding economic, social, and environmental consequences for the Northwest. Salmon fishing is big business in the region, worth, until the recent falloff in the fish population, about $1 billion a year in revenue and more than sixty thousand jobs. But in the past few years, as many as one-third of northwestern fishermen have lost their livelihood. Meanwhile, for the Indians of the area, the demise of the salmon is heartbreaking. The fish is a staple of their diet and a totem of the natural and abiding rhythms of life. The Indians traditionally believed that salmon were people who live in villages under the sea and disguise themselves as fish to visit and feed human beings, and they still hold religious services when the few salmon reappear. During these ceremonies, they put the bones of the first fish they catch back into the water so it can be reborn, propagate, and return again.

Perhaps the most enduring ramification of the salmon's decline is its impact on the ecosystems of the rivers where they live part of their lives. The environmental importance of the curious cycle of Pacific salmon hasn't been clearly understood until the past few decades. In contrast to Atlantic salmon, which give birth and return to the ocean many times, Pacific salmon die as they spawn. The rivers from which Pacific salmon come are generally low in nutrients. While in the ocean, salmon eat food—bugs, fish, and plant life—that is rich in proteins and minerals. When they die upstream, their decaying carcasses become a steady source of high-powered nutrients for the river's plant life and even the salmon's own offspring.

As it must for every animal on the endangered species list, the federal government is required to put together a plan to rescue the Pacific salmon. But a number of powerful conflicting interests in the region are making saving the fish virtually impossible. One suggestion floated early on was simply to breach, or tear down, many of the dams and give the salmon a freer course between the rivers and the ocean. Indians, fishermen, and environmentalists all backed the idea. And even the electric utilities—which are discovering that as the dams age, they're costing quite a bit more to maintain and are not as profitable as they once were—are willing to part with some of the structures. But if the dams come down, flooding would inevitably occur; developers and agribusinesses, supported by lawmakers including Washington's former Republican senator Slade Gorton, refused to consider giving up any of their land to those floods. Meanwhile, Indians and fishermen have consistently rejected any plans that significantly reduce the number of fish that can be caught on the rivers each year.

Unable to forge an agreement that would satisfy all of these conflicting interest groups and unwilling to impose one at the risk of alienating any of them—they all have vocal supporters in Washington and can rally votes in elections—the federal government has given up. Instead of developing a plan for saving the salmon, the government has decided to essentially do nothing for another few years while the possibility of removing dams, combined with other solutions, is explored a bit more.

For the Snake, Columbia, and Okanogan rivers and Salmon Creek, among many other waterways in Washington state, this stasis may be devastating. The number of salmon will surely continue to dwindle, and the loss of the fish that exemplifies the Northwest will have further ripple effects. By refusing to budge, all sides in the dispute are willing a potentially painful outcome. For the Skagit River, though, the result may be completely different.

Skagit Valley, where the river cuts through as it nears Puget Sound, is about sixty miles north of Seattle but looks as if it could be in another country, or at least another state. Its small towns with wooden storefronts have a western feel like that of the Dakotas or Montana, and the Skagit River itself is much less developed than the other salmon rivers, which are generally much larger waterways. Moreover, Skagit Valley shares with places like Montana an independent streak and a simmering distrust of the federal government.

This distinction is neither artificial nor toothless. In 2000, I spoke to at least two dozen people in Skagit Valley, all of whom described watching in horror as the federal government mulled over the future of their river. That the government might dictate the future of their valley and the course of the Skagit was a prospect they hadn't taken seriously before, but it had since become uppermost on their minds. Subsequently, the different factions in the Skagit-area dispute—developers, farmers, Indians, environmentalists, and fishermen—formed a loose-knit organization called the Skagit Watershed Council. Its goal is to resolve the dispute over the salmon, without help from the federal government.

None of these people are particularly fond of each other. Developers told me that the Indians and environmentalists are "antiquated anachronisms," as one put it: "They have their little mantras, and all I can say to them is grow up." One Indian I talked to spit on the ground when describing his feelings about the farmers who "steal away each year more land near Puget Sound for crops, leaving no place for the salmon to mature."

Still, despite these differences, the council is achieving small victories. In almost a mirror image of Gorbachev's approach, the group's leader, Shirley Solomon, a tall, tireless woman of South African descent, makes sure that all sides in the dispute meet at least every month, usually to present some bit of information they were asked to research.

"I want all of us to begin to see each other as human beings, not representatives of a position," Solomon told me in the Watershed Council's spartan offices on the second floor of an old office building in downtown Monroe, Washington. The furnishings are card tables instead of desks; maps of the river, with its dams clearly marked, cover one wall; and a sculpted salmon in classic curled pose hangs on the other.

The group's largest accomplishment has been to agree on a statement of purpose that the salmon must be saved from extinction; it means too much to Skagit Valley. Most important, the Watershed Council has been able to follow up that statement by lining up a series of state grants to reclaim land that salmon use for habitat and to unblock some of the Skagit's course so that more adult salmon can return to their spawning grounds.

Solomon, an ardent environmentalist—she's been associated for years with a salmon protection organization called Long Live the Kings—admits that it's highly unlikely that the council members will ever agree that the Skagit dams should be breached—even though the utilities, like Seattle Light, don't care if they are. But these days, she says, she tries to put old opinions—like the necessity of destroying dams—out of her mind, because the purpose of the Watershed Council is to teach its members to not be so inflexible or assume that everything they once believed is true.

Solomon explained: "This is how we see the water conflict in Skagit Valley now. Typically, we tend to think in black and white, and all our lives are played out in shades of gray. So you settle for compromise all the time. When in actual fact if you get a partnership and a collaboration going—as we have—new colors arise. Not black or white or gray. And suddenly instead of compromising, you're actually sharing the same beliefs, not some watered-down concession." I found this a refreshing point of view, a rare one in water conflicts, since usually the parties' failure to let themselves get even as far as gray dooms any attempts to resolve the dispute.

Achievements like those in Skagit Valley and on the banks of the Pilcomayo River amount to much more than the sum of their telling. In deciding to stop the damage to water—and to those who live in it and drink it—and to end the local water wars that have caused so much of the harm, both of these regions have made a significant statement: the idea that anyone wins when water is a weapon is foolish and dangerous, and one that people have toyed with for too long to consistently losing results. Hundreds of other regions are still ignoring this lesson, with often tragic consequences. That record—a couple of victories against dozens of losses—doesn't just amount to abject failure; it also, as Madeleine Albright warned in her Earth Day speech, raises the far deeper question of whether we actually know how to survive or only know how to misuse the sole natural element upon which our survival most depends.

8

Planet Water

PROBABLY THE ONLY positive thing that can be said about the water crisis is that it's provided a lot of work for water scientists and engineers. Water systems around the world may be drying up because of polluted and diverted rivers or—where the water supply was adequate in the first place—crumbling as local governments can't meet the cost of supporting the infrastructure. But numerous researchers are receiving hundreds of millions of dollars in funding each year to explore water delivery alternatives and study even the most obscure aspects of water's life cycle. We've even sent probes to Mars to see if there's water forming there.

One might well argue that instead of spending billions on research, we should just use the money to supply water to those that need it. But that's not so easy anymore, not now. After all, the World Bank and other international lending agencies have poured much more money than was earmarked for water-related scientific investigations to build huge projects around the world that ostensibly were supposed to deliver water for industry, irrigation, and residential use in some of the poorest regions in the world. But the result of these efforts—and other overscaled water ventures, like bringing the Colorado River to southern California or turning the Nile into the world's biggest reservoir—has been that less water is available than ever before, just when we need more of it. Put plainly, we don't really know how to manage water anymore; we've made all the wrong choices and have all the wrong ideas about what water is and who we are in relation to it. Scientific investigations of water, both practical and theoretical research, hold out the hope of at least getting us back on the right track.

One area of research that has received a lot of attention in the past five years or so from the scientific community is already producing significant benefits. It's the technique of desalination, which removes salt from ocean water, making it usable for freshwater purposes. In some ways, the idea is an old one. Aristotle wrote in the fourth century B.C. that "saltwater, when it turns into vapor, becomes sweet and the vapor does not form saltwater again when it condenses." But for Aristotle and others who later promoted the idea of desalination, the aim was to get the salt from the water, not the water from the salt. Pure salt was of immense value; water seemed to be everywhere.

The situation is quite different now. With so many poverty-stricken water-scarce regions and with so many populations exploding where they shouldn't—in the desert, for instance—the notion that seawater, which amounts to 97 percent of all the water on Earth, could be used as drinking water is extremely attractive. In the twenty-first century, water is definitely worth more than salt. As an alternative to shipping water over seas from areas with a surplus to places that are begging for it, it seems logical to use the oceans themselves, which are so abundant that we'll never run out of them no matter how large the Earth's population becomes, as a water supply. In other words, we could cut out the middleman—or in this case, drink the middleman.

As recently as the 1960s, desalination research received strong backing from government agencies, who mostly wanted to use the technology to supply water for remote military operations. But despite the money spent on it and the research put into it, desalination was difficult to refine with the scientific techniques of the time. Consequently, it remained extremely expensive and not particularly effective. Many desalination systems could change only about 25 percent of salt water into freshwater, which made them too inefficient for their high price tags. Consequently, few in the water scientific community thought desalination was worth exploring.

That view shifted as the water crisis worsened. In 1996 former Senator Paul Simon, one of the few politicians to call attention to the worldwide water scarcity, convinced lawmakers to back a research program that would investigate ways to improve desalination technology. Though Simon's bill provided only limited direct financial support for water scientists, it required matching funds from the private sector. That, at least, encouraged U.S. companies as well as international funding agencies to take another look at desalination. And in the past five years, as the amount of money poured into desalination research has rapidly increased, the technology has finally been perfected to the point that in some regions, particularly in parts of the Middle East, it supplies more than 50 percent of local water.

Two techniques for desalting ocean water have produced the most significant recent breakthroughs. The first is reverse osmosis, in which huge amounts of pressure are used to pass water through a semipermeable membrane, which filters out the salts. A variation known as electrodialysis uses electrical currents to move salt ions through the same kind of membrane employed in reverse osmosis, leaving freshwater behind. The noteworthy engineering advance for both approaches was the development of exceptionally sophisticated membranes that permit passage only of water with a specific chemical makeup from one side of the desalination system to the other.

The second desalination technique that is becoming more popular is called thermal distillation. It's particularly creative because it mimics a part of the hydrological cycle itself: it uses solar heat to warm salt water enough that water vapor is produced—very similar to that which rises out of the ocean into the atmosphere. This saltless vapor is then transported through a porous thin film, after which it's cooled and turned into liquid freshwater.

Not surprisingly, much of the initial use of desalination has been in desert regions like the Sinai and Saudi Arabia; 60 percent of all desalination plants are, in fact, located in the Middle East. But it's also being im-plemented in more tropical areas like Trinidad, Tobago, and Tampa Bay. There are twelve thousand desalination plants worldwide, producing more than 5 billion gallons of water each day. That sounds like a lot, but it's ac-tually only a tiny amount—just two-thousandths of total world freshwater use. Or put another way, the total amount of desalinated water produced annually is about as much as the entire world consumes in a little more than a half a day.

One obstacle to desalination is that it's still expensive, though less than what it used to be: it costs $2,000 or more per acre-foot, while in many re-gions water from lakes and rivers costs only half that. For this reason, it's being adopted so far only in countries that are near an ocean but have very little interior water and that have enough money, or at least backing from international agencies, to support such a costly operation.

The project in Tampa Bay, on Florida's west coast, is a bit of an anom-aly. Much of the United States has resisted desalination, because Ameri-can consumers are reluctant to pay higher water bills, even when they live in areas that are water scarce. In southern California, for instance, which has one of the world's most bitterly dry deserts and lies next to one of the world's biggest oceans, desalination is being considered very gingerly. It has been deployed or proposed in only a few communities, while other less ex-pensive but potentially damaging approaches like tapping desert aquifers are eagerly embraced. In Tampa Bay, which like the rest of Florida is running out of water, the choice of desalination was more the result of des-peration than innovation. Tampa Bay Water, which serves Tampa, St. Peters-burg, and New Port Richey, expects the plant, when fully operational at the end of 2002, to produce 25 million gallons of drinking water daily, pro-viding 10 percent of the area's needs. The increased cost will be passed along to consumers—a hopeful sign, in the eyes of desalination industry executives.

"Desalination has never had a lot of public support in the United States," Fared Salem, DuPont Company's desalination manager, told me. "But maybe when people see how well Tampa works and how limited the financial pain is, there will be renewed interest in a technology that can really address the twenty-first century's water problems."

By contrast, extremely water-scarce but oil-rich nations like Saudi Arabia are concerned less about cost than about availability. The Saudi government is footing the bill for an ambitious $2 billion desalination plant that, combined with the nation's already significant desalination capacity, will anchor a water supply system that covers some two thousand miles of pipeline and more than a hundred reservoirs—all of it supplied by seawater. Currently, desalted water provides 70 percent of Saudi Arabia's drinking water needs. After the desalination expansion is completed, virtually all of the nation's water will come from this technology.

The price of desalination will have to drop quite a bit before many of the poorer nations can afford to use it. In fact, it may take years before it's an appropriate option for them. But some water experts hope that perfecting desalination now in richer countries will eventually make the technology cheap enough that emerging countries can install it, especially as international agencies like the World Bank respond to pressure to change their funding policies from providing money for mostly large dam projects to backing safer water supply alternatives like desalination.

Desalination has given rise to a philosophical split in the scientific community. The more practical researchers, who are working on desalination technologies, view it as their responsibility to help find immediate and concrete solutions to the water crisis. Merely studying the problem, but not doing anything to solve it, they feel, would be like standing in the middle of a raging fire and analyzing a flame.

But a persuasive theoretical wing among water scientists is concerned that desalination technologies, and perhaps other short-term fixes as well, can actually devalue water. In their assessment, desalination encourages the attitude that the seas, like the rivers before them, are something that we can control instead of respect. While there's no imminent danger of running out of ocean water, we've already harmed the coral reefs, they point out, the most diverse habitat on Earth, which grow in warm tropical and subtropical oceans and are made up of the skeletons of millions of tiny animals called coral polyps. Eight hundred species of reef-building coral have been identified in these incredibly colorful and disparate environments that provide habitat and nutrients for about four thousand species of fish, many of which feed a large part of the world's population—one billion people in Asia alone.

More than a quarter of the world's coral reefs have already been destroyed, however, mostly by pollution or temperature changes brought on by rivers emptying into them that are hotter than they should be due to dams and pipelines upstream. Another 60 percent of coral reefs are threatened and could be lost in the next fifty years. Some water scientists fear that tampering with the oceans through widespread desalination could lead to subtle changes in seawater, such as minute increases in salinity, that make it impossible for certain fish to survive. So far there's been no evidence to suggest that this will actually occur, but considering our track record with water management, concerns of this sort cannot be discounted. Worries like these, in a sense, underlie the research of theoretical water scientists, who strive to shed light on the secrets of water, to explain the vast amounts that we don't know about it, so that people can observe how diverse, miraculous, and essential it is—and how much of our existence we owe to it. Maybe through greater understanding, their thinking goes, we will fully recognize water's radiance and fragility and begin to use it as a resource in ways that also ensure its preservation.

"It's critical that we understand the place of our lives in the universe and in nature," Penelope Boston, a microbiologist who founded a research firm called Complex Systems, told me. Boston has participated in dozens of explorations to investigate primitive creatures and ecosystems that thrive on the ocean floor, and she has recently expanded her area of interest to the search for life on other planets. "To the extent that we simply behave as organisms in the test tube of the Earth without understanding what we are doing, we suffer the consequences of organisms who don't understand what they are doing. Ultimately, the more we know about things, the more prepared we are to try to ameliorate the consequences of our actions on Earth."

In all of my research, nothing more vividly demonstrates the power of this point of view and the magnitude of water's impact on Earth—its fundamental role in the planet's creation and its eternal regeneration—than the experiences of a group of oceanographers who observed the spontaneous emergence of life in the place that it likely first occurred: hot pockets of gas, known as hydrothermal vents, at the bottom of the ocean. The past three decades or so have seen a series of hydrothermal vent discoveries, but the one that most captured the attention of scientists occurred just before midnight on December 3, 2000, in the Mid-Atlantic Ridge off the East Coast of the United States.

The *Atlantis*, a 274-foot research vessel out of Woods Hole Oceanographic Institution in Woods Hole, Massachusetts, was nearing the end of its four-week exploration of the Mid-Atlantic Ridge, which at six thousand miles wide is one of the Earth's largest undersea mountain ranges. The boat's crew members were lingering on the ship's upper deck, chatting about the next day's activities. The *Atlantis* was towing a tethered, deep-sea camera that sent back real-time video along a fiber optic cable, and a digital camera that flashed images of microscopic activity to the control room every fifteen seconds.

Suddenly, the cameras began picking up unusual sightings: the monitors were filled with the shapes of what one crew member described as "amazing white structures." They were 180-foot-high towering spires, thirty-feet across at their tops, many of them sprouting delicate, bowl-shaped flanges that feathered out from their sides. At first, the scientists on board thought they were coral, but no living organisms were immediately visible. If they were hydrothermal vent fields, the crew members thought, they were the largest and strangest ones anyone had ever come across. "They didn't look like anything I'd ever seen before," recalls Deborah Kelley, one of the lead researchers on the *Atlantis* and a professor of marine geology and geophysics at the University of Washington.

The first hydrothermal vents, which are actually caused by volcanic activity on the Earth's crust, were discovered in the 1970s, and many scientists were immediately convinced that they had found the place where life begins. These seabed hot springs generally look like columns and are characterized by fluid venting from the chimney's sides. The molten liquid creates unusual mushroom-shaped growths—flanges—that jut outward and trap more and more of the heated fluids and dissolved gases.

Lurking in these writhing ecosystems are communities of Archaea, the infinitesimally small bacteria at the very beginning of the food chain. Archaea can spawn and thrive in the buried depths of the ocean and in the crags and gullies of hydrothermal vents because they don't need sun or photosynthesis for survival. Instead, they use sulfide, which is plentiful in the vents, to create food, in a process called chemosynthesis. And they eat nutrients from seabed rocks, which are transferred to them by the water. Ever since the discovery of Archaea in the late 1970s, more than three hundred species of vent animals higher up the food chain that feed on Archaea have been identified in hydrothermal vent habitats—exotica like the red-tipped, white Vestimentiferan tubeworm; the giant Vescomyid clam; the Bathymodiolid mussel; and the deep-water skate.

But what was different about the hydrothermal vent that suddenly appeared on *Atlantis*'s monitors was the absence of visible life. There were no tubeworms, no clams or mussels, only the Archaea and an incredible amount of volcanic activity. As the crew moved the cameras back and forth to examine different parts of the hydrothermal vent—which by now was dubbed the Lost City, because of its primordial appearance—their excitement mounted. Unlike other vents that looked more mature, the Lost City had fresh white-colored deposits in its youngest areas, and hot viscous fluids discernibly streaming up the edges of its pillars, as well as brand-new crystals forming on its edges. It was possible, the crew thought, that with this feverish explosion of spewing hot liquids, crystallizing to extend the walls of the vast ziggurat, and with the emergence of the tiny Archaea in the middle of the blast furnace, they were actually watching life begin.

"These structures were forming right before our eyes by the precipitation of minerals from fluids that derive solely from chemical reactions, not from anything else," Kelley told me. "This was a gigantic surprise."

The next day Kelley, Jeffrey Karson (Kelley's research partner aboard the *Atlantis* and head of the geology department at Duke University), and a pilot squeezed into a small submarine and dove deep into the ocean to get a closer look at the Lost City. As they neared it, they turned on the high-beam klieg lights. Kelley and Karson were spellbound, staring through the submarine's porthole, as they slid down the south side of the face of the vent's scarp and then rose up until they came to the edge of the tallest structure. It seemed to go on endlessly. "It was incredible," Karson says. "Every time we found a new structure, it was like, my God, this field is just getting bigger and bigger and bigger."

No science fiction movie about searching for a lost ark or buried treasure had prepared them for what they saw. There was movement everywhere, constant, undulating movement. It seemed that the Lost City was made up of thousands and thousands of kinetic parts each creating ecosystem after ecosystem, rhythmically pulsating in concert as if they were enlivened by a drumbeat that no one else could hear, building a steamy hothouse from which life could emerge. With robotic arms, the submarine's crew grabbed hundreds of samples of the Lost City to take back to land. Then after four hours at the hydrothermal vent, the crew had to return to the home vessel. It was almost impossible to pry themselves away. Speaking about the exhilarating experience months later, Kelley sounded exhausted, as if just recalling it overpowered her again. "I've struggled for words to describe what I saw," she says. "They don't exist."

It will take the *Atlantis* oceanographers some time before they figure out what they found in the Lost City. Researchers at Woods Hole are testing the specimens they brought back and mulling over the photographs, videos, logs, and researchers' diaries. Before anything is concluded with certainty, numerous teams of other water scientists, representing a vast range of organizations, are likely to take further exploration voyages to the Lost City. The *Atlantis* crew is still a bit shell-shocked by what they saw. Lingering with them most of all is the ferociousness of the formative activity they witnessed firsthand—the establishment of mountains out of the planet's core, the spontaneous blast of chemical reactions, the uncoupling and emergence of primitive plant and animal life, the belching screams of volcanic eruptions, all of which, they know, continues to play out at the bottom of the ocean day and night. It's impossible to forget the sight of the planet in a burst of anger and frenzy recreating itself over and over in the deepest ocean, where the Earth is most at home.

In a strange twist of fate, the other recent, significant, purely theoretical scientific discovery involving water occurred about as far from the Lost City as most of us can fathom—millions of miles away on Mars. For years, scientists have hoped to discover water on Mars, mainly because the planet, the fourth from the sun, is the only one besides Earth that has had the narrow temperature band in which water can exist in all three states: solid, liquid, and vapor. Martian meteorites landing on Earth have had small concentrations of water, and thus some scientists have been led to speculate that water may have actually originated on Mars. The problem is that these meteorites tell us very little about conditions on Mars today, because they were ejected from the planet nearly 200 million years ago. With the current climate, though, if there were liquid water on Mars, it would have to be underground, probably in environments similar to the ocean's hydrothermal vents. The temperature on the planet's surface has dropped to an average of minus nineteen degrees Fahrenheit; consequently, only ice could be found there. But Mars is covered with deep channels and canyons and what appear to be ancient lake shorelines—ample evidence that billions of years ago liquid water may have flowed there. Because water is the fundamental element for creating and sustaining life, finding it on Mars in any form would present the real possibility that some type of life either exists on the planet now, or did once before, or will someday.

On January 4, 1999, the Mars Polar Lander lifted off atop a Delta II rocket at Cape Canaveral toward the Red Planet's south pole. The plan was for it to complete its 470-million-mile journey on December 3 of that year—late spring on Mars, when the sun never sets. The spacecraft would be the first ever to land on the polar region of another planet.

Viewed from above, the Lander looks like a giant monarch butterfly, with its jutting, antennalike instrumentation and its spread-out solar-panel wings.

Five minutes before entering the Martian atmosphere, two basketball-size probes were programmed to detach from the main spacecraft and crash into the planet at four hundred miles per hour, penetrating three feet into the Martian soil in the hope of finding frozen water there. At around the same time, a camera—the Mars Descent Imager (MARDI)—mounted on the bottom of the Lander was prepared to take high-resolution photographs as the spacecraft made its final approach. Upon landing, the Lander's six-and-a-half-foot-long robotic arm would scoop a few inches of soil for chemical analysis on Earth, and its instruments would track weather conditions. A microphone attached to the Lander would be the first to actually pick up sounds from another planet. The mission's price tag was relatively cheap, about $150 million—part of a calculated ten-year NASA strategy: one mission to Mars every twenty-six months.

"We think there will be some frost, if not ice, on the layered terrain," Ed Weiler, head of the National Aeronautic and Space Administration's space science division, said soon after the Lander took off. "But we really don't know what we are going to see. You've got to follow the water if you're looking for life."

Unfortunately, the 630-pound Lander never delivered any answers. The spacecraft was ready to float for its final descent when—twelve minutes shy of its scheduled touchdown—it disappeared. In a report prepared months later, NASA blamed the mission's failure on spurious signals that were emitted when the Lander's legs were dropped during descent, which gave a false indication that the spacecraft had landed. As a result, the engines shut down prematurely and the Lander crashed into the planet's surface.

A frustrated Weiler told the press soon after the Lander vanished, "It doesn't take a rocket scientist to ask the question, maybe we had too lofty goals, maybe we're being too aggressive."

Privately, though, Weiler said that even if the engineering was faulty, neither the aim of the mission nor its timing was a mistake. The search for water on the Red Planet is "an absolutely essential exploration," Weiler said—and that's what was most disturbing about the loss of the spacecraft.

Little more than six months later, however, something completely unexpected occurred. At a June 22, 2000, press conference, NASA announced that it had, indeed, found evidence of water on Mars. Definitive proof, NASA said, was discovered north of a section of the planet called Vales Mariners—not far from where the Viking I Mars exploration spacecraft had landed in 1976.

It turns out that the same company that had designed the MARDI camera on the Polar Lander had attached a camera called MARCI (Mars Color Imager) to the Mars Global Surveyor, which was launched in 1997 to orbit the planet and produce maps of its surface until January 2000. After that—indeed, just about the time the Polar Lander was supposed to arrive—the Global Surveyor was slated to be only a communications satellite, relaying data to Earth from other spacecraft that were sent to Mars. Even after the Global Surveyor was mostly finished with its work, MARCI continued to take pictures of the planet as the spacecraft orbited it, from a distance of 235 miles.

The photographs supposedly showing water on Mars are incredibly detailed, even if to the naked eye it's not clear that they do show water. There are large flow channels coming out of the cracked Martian terrain, resembling winding rivers on Earth, estimated at more than sixty miles wide and more than a thousand miles long. And there are what looks like eroded craters with teardrop-shaped tails, scour marks, and islands.

"We think we are seeing evidence of a groundwater supply, similar to an aquifer," says Michael Malin, president and chief scientist of Malin Space Science Systems, which designed the MARDI and MARCI cameras.

Malin's theory is this: on the coldest slope of Mars facing away from the sun, underground water is freezing and forming subterranean ice dams. When the water pressure behind the dams builds up enough, they rupture. An explosive slurry of water, ice, and rocks cuts a swath beneath the surface and then carves the gullies and channels that can be seen in the photographs.

"Had this been on Earth, there would be no question water was associated with these weeping layers," Malin says, displaying a dramatic photograph of a gully-carved cliff wall the color of creamed mocha. "The features appear to be so young that they might be forming today."

Not everyone subscribes to Malin's hypothesis. It's a bit too tidy and difficult to prove or disprove, because it's based on underground, unseen activity and photographs taken hundreds of miles away that show no clear evidence of real water. In fact, some skeptics go so far as to say that NASA essentially jogged Malin and his "water" photos out to build some positive public relations in the wake of the Polar Landing debacle and other similar mishaps—notably, the 1993 loss of the $1 billion Mars Observer—that have plagued missions to the planet.

Even if Malin's ideas are correct, to actually extract and analyze some of this "water" just to find out whether it is indeed water will likely cost billions of dollars. Which gets back to the question that I found impossible to neglect: Wouldn't that money be better spent on practical water research—projects that are designed solely to solve the water shortage, projects that actually provide water for even one person who at this moment is dying of thirst? In other words, isn't it too late in the water crisis for soft science?

Listening to the accounts and looking at the videos and photos of life emerging in an undersea thermal eruption and of what seemed to be an exclusively Earthbound element possibly flowing on Mars, I was as wide-eyed as anyone else. They're absolutely remarkable stories. But as hard as I tried, I couldn't see clearly the principal distinction between our severe mismanagement of the little water we've been given into a serious water crisis and researchers' visits to the most exotic of all water ecosystems to do nothing but probe, remove samples, and theorize. Their obvious differences aside, neither effort provides a glass of water to the billions of people on Earth who face extreme suffering—and dripless taps.

At the beginning of the 1960s, John Kennedy threw down the challenge to achieve a moon landing by the end of the decade. American scientists and engineers responded eagerly. It may have seemed like a critical goal at the time—and an almost impossible one to attain—but solving the water crisis we face now is much more essential to our survival and will be much more difficult to accomplish. Whether we accomplish it will decide our destiny. We will need all the scientific know-how we can get.

Ending: Salvation

Now, in the waning of light,
I rock with the motion of morning;
In the cradle of all that is,
I'm lulled into half-sleep
By the lapping of water.

THEODORE ROETHKE,
"MEDITATION AT OYSTER RIVER"

IN JUNE 1968, I was at Resurrection City, a temporary town of tents erected in front of the Lincoln Memorial by thousands of America's poor, who had been urged to come there by Martin Luther King, Jr., before his murder just a few months earlier. People, white and black, had traveled to Washington, D.C., from every forgotten location of the country, arriving in aging pickup trucks with broken slat wood and chicken feathers on the back bed, old station wagons that comfortably seated eight but were now crammed with a dozen passengers, and even antebellum mule carts. They were there to demand jobs and a living wage.

It was one of the most remarkable places I've ever been. At night, kerosene lamps and assorted bonfires lit up the makeshift streets that ran through the grounds, providing a haunting Old World backdrop that made it seem like a simpler time even though the late 1960s were anything but simple. Music—guitars, dulcimers, banjos, harmonicas, and a capella voices hitting six- and eight-part harmonies—echoed from every corner. As a teenager from New York City, who had become a guitarist a few years earlier and was spellbound by blues and deep country, I couldn't believe my good fortune to be among people who had learned these songs because they were born where they were sung and never had to study them the way I did.

But sadness and even bitterness also permeated Resurrection City. Each face held hardly a hint of a smile, ever. Few people talked to anyone besides their immediate families and the neighbors they had come with. These were tired, angry people, who wanted more from life than they had gotten so far. It wasn't to be—not here, and perhaps, for most of them, not in this existence at all. President Johnson refused to even recognize the legitimacy of Resurrection City or speak with its leaders. He called the tent town a blight, especially after rainstorms pounded it and turned it into a mud-soaked health hazard. He said he'd never negotiate with troublemakers.

In the middle of this misery, though, something unexpected happened that lifted the dour mood from some Resurrection City residents, at least for a few moments. It was near midnight, and a revival meeting was just ending in one of the big tents, when the Reverend Jesse Jackson led hundreds of congregants, walking in a straight line, some holding tightly to crosses, to the reflecting pool that separates the Lincoln Memorial from the Washington Monument. One after another, they stepped into the water, almost mechanically. Jackson baptized each of them. And as they emerged out of his arms, they began to dance wildly in the water, kicking up their legs, splashing each other and the air without a touch of self-consciousness. Their downcast expressions changed to joy, and their bodies seemed unchained, and the cheerlessness of their lives was forgotten, while the majestic and pure white reflection of Lincoln, seated on his huge marble throne, quivered eerily beneath them in the black pool. Just days later, Washington police tore down Resurrection City and its dwellers dispersed.

I was haunted by that incident and still often consider with amazement the extraordinary uplifting effect that that water had on their downcast spirits. These people had traveled a long way, hoping for a miracle to bring them some relief from the grind of daily poverty. They got none—except when they felt the fresh chilly spray of "Lincoln's water," as a baptized man called it, washing over their bodies. At that moment, they were transported into another skin and a better life. One woman climbing out of the pool, breathless and flush, said to me, "We did it this time. I think we really crossed the Jordan River."

It's fascinating how enduring a metaphor the Jordan River is, I thought much later, as I recalled her words. For more than three thousand years, the Jordan has been the symbolic boundary between persecution and freedom. Its waters have stood for salvation. In fact, the Jordan remains in our collective memory oddly pure and comforting, especially since this helix-shaped 135-mile river that descends out of the snows of Syria's Mount Hermon and winds circuitously to the Dead Sea is now so polluted and drained by misuse and a tug-of-war among Middle Eastern nations that it is as damaged as almost every other river in the world. In truth, it's now just another distressing manifestation of the water crisis.

How strange that, after so many centuries, we should still cherish the idealized image of the Jordan and ignore the river's currently worsening darker reality. Strange, yes, but atypical, no. Think of how many times you've stood on the sand staring out at the ocean, hypnotized by the foam shagging back and forth over your feet, or the moments when you dipped your toes into a babbling brook, watching tiny fish curl around your skin— just touching the soft edge of nature. And then consider how carelessly we've treated water, probably even the water lapping against your skin— and how infrequently we even consider what we've done.

Researching this book has been as inspiring an experience for me as the Resurrection City baptism in the reflecting pool. At the start of my investigation, I, like most Americans, didn't even know—and probably wouldn't have thought it mattered—that three-quarters of the water we use in the United States flushes down the toilet or runs down the bathroom drain. I never thought twice about keeping a faucet on full tilt, while I walked away to do something else, hoping it would be warm enough to use by the time I returned. The idea that front-loading washing machines require 40 percent less water than top-loading washers wasn't a concern of mine. Water scarcity, I believed, was an anomaly, occurring in other places where people die all the time of awful environmental conditions; I assumed that floods were controllable in developed nations, and that the

United States had so much water that no matter how many dams we built, our supply wouldn't run out. In short, the world's water—in crisis or not—was someone else's concern and had nothing to do with the water I used.

I'm a bit embarrassed now by my naiveté and ignorance. Most eye-opening to me was finding out that the water crisis actually exists in every nation. All of us are running out of freshwater; some places, underdeveloped countries as well as areas in the United States, never had any water in the first place. Everyone has mismanaged the limited supply we were given, and now we're living on borrowed time.

The potential solutions to the water crisis that we have considered may hold some cause for optimism. The new technologies of water transfer and desalination are promising. Privatizing water systems with local citizen participation could add a profit motive to delivering water that will improve its quality, especially in poor countries where governments have failed miserably. The idea of applying true market pricing to water in wealthier countries to encourage conservation and then using that money to subsidize water systems in nations that can't afford to develop their own is gaining support even from environmentalists. And lending agencies are rethinking their support for the big dams and diversion projects that have destroyed communities and polluted water supplies; instead, they may begin to offer funding for local development projects that could supply water for small groups of residents with less potential damage to the water supply.

None of these efforts will work, though, if we don't begin to realize that water is so vital to the planet that when we damage water supplies anywhere, it's as if it's happening right at our own tap. The world for all of us has shrunk, and the Jordan River—the one that exists today and not the idealized river that none of us has actually seen—has moved next door.

Some twenty years ago Pete Seeger, one of the wisest men I've ever known, tried to teach me this, but I missed the point completely. I was a volunteer on the *Clearwater,* the early-1800s replica single-masted boat that Seeger had built to sail the Hudson River and work with local residents to develop programs for cleaning its polluted waters. When he founded the *Clearwater* project, Seeger was a longtime folksinger, a contemporary of Woody Guthrie, who had fought tirelessly in some of the most difficult social and political struggles in U.S. history—for worker and civil rights, against the war in Vietnam and the blacklist.

I first met Seeger when I was just becoming a musician. I performed benefits with him and was inspired by the broad moral positions he took and his refusal to compromise on any of them. So I was surprised when he founded the *Clearwater* project and downplayed his involvement in more weighty issues, to focus instead on a single river. I asked him how he could possibly think that cleaning up the Hudson even mattered compared to fighting for civil rights and living wages. Perhaps he was making a mistake, and misdirecting his priorities, I suggested. Seeger didn't answer right away. He seemed to be thinking about what I had said. Then he looked at me with the squint and tiny smile that I knew meant, *Listen carefully—I'm about to teach you something.*

"You have it backward," he said. "If you save one drop of water, you've saved the world."

Notes

IN RESEARCHING THIS BOOK, over a period of a year and a half I read hundreds of articles, reports, and books; attended any number of seminars and speeches; and browsed countless websites on the subject of water. I employed researchers for help in sifting through the huge amount of material that is available about water. In some instances, these researchers provided anecdotes and conducted interviews that are included in the book. In all cases, I painstakingly fact-checked their work. For consistency, I maintained the first person voice in the book, even in the few times when a researcher did the initial reporting. For those interested in reading more about water and the water crisis, I have compiled a list of the most essential and intriguing material I uncovered and the websites I came to rely on. This is a foundation for further research, a starting point for exploring a huge, often complex topic.

BOOKS

Ball, Philip. *Life's Matrix: A Biography of Water.* New York: Farrar, Straus & Giroux, 2000.

De Villiers, Marq. *Water: The Fate of Our Most Precious Resource*. Boston: Houghton Mifflin, 2000.

Douglas, Marjory Stoneman. *The Everglades: River of Grass* (50th anniversary edition). Sarasota, FL: Pineapple Press, 1997.

Gardner, Jason, ed. *Sacred Earth: Writers on Nature & Spirit*. Novato, CA: New World Library, 1998.

Gleick, Peter H. *The World's Water: The Biennial Report on Freshwater Resources, 1998–1999*. Washington, DC: Island Press, 1998.

———. *The World's Water: The Biennial Report on Freshwater Resources, 2000–2001*. Washington, DC: Island Press, 2000.

Hardin, Blaine. *A River Lost: The Life and Death of the Columbia*. New York: Norton, 1996.

Reisner, Marc. *Cadillac Desert: The American West and Its Disappearing Water*. New York: Penguin, 1993.

Shabecoff, Philip. *Earth Rising: American Environmentalism in the 21st Century*. Washington, DC: Island Press, 2000.

Sharp, Robert P., and Allen F. Glazner. *Geology Underfoot in Death Valley and Owens Valley*. Missoula, MT: Mountain Press, 1997.

Snyder, Gary. *A Place in Space: Ethics, Aesthetics, and Watersheds*. Washington, DC: Counterpoint Press, 1995.

Yergin, Daniel, and Joseph Stanislaw. *The Commanding Heights: The Battle Between Government and the Marketplace That Is Remaking the Modern World*. New York: Simon & Schuster, 1998.

ARTICLES, REPORTS, AND PRESENTATIONS

Albright, Madeleine K. "An Alliance for Global Water Security in the 21st Century." Speech presented at National Defense University, Washington, D.C., in recognition of Earth Day, April 10, 2000.

Barlow, Maude. "Blue Gold: The Global Water Crisis and the Commodification of the World's Water Supply." Report issued by the International Forum on Globalization, June 1999.

———. "Commodification of Water—Wrong Prescription." Presentation to the 10th Stockholm Water Symposium, August 17, 2000.

———. "The World Bank Must Realize Water Is a Basic Human Right." *The Globe and Mail* (Toronto), May 9, 2000.

Biddle, RiShawn. "Cadiz in the Dumps." *Forbes,* July 3, 2000.

Broad, William J. "Reports of a Dead Mars Are Greatly Exaggerated; Red Sands Show Marks of Recent Water and Lava." *The New York Times,* July 25, 2000.

Caplinger, Michael. "Life on Mars." Malin Space Science Systems, msss.com, April 1995.

"Chemical Analysis of Six Martian Meteorites Indicates That Water Was Once Abundant on Mars." nasa.gov, January 22, 2001.

Clifford, Frank, and Tony Perry. "Desert Water Entrepreneur Closely Tied to Governor." *Los Angeles Times,* April 16, 2000.

Darwish, Adel. "The Next Major Conflict in the Middle East? Water Wars." Presentation at the Geneva Conference on Environment and Quality of Life, June 1994.

Fernandez, José W. "Providing Water in Latin America." *New York Law Journal,* April 27, 1998.

Firestone, David. "Looking Past Georgia's Drought, but Still Seeing Brown." *The New York Times,* March 5, 2000.

Fountain, John W. "Public Lives: Leading a Grateful Fargo in a New Fight Against Floods." *The New York Times,* April 14, 2001.

Glanz, James. "In Defense of Defenseless River Towns." *The New York Times,* April 29, 2001.

Grunwald, Michael. "Rivers of No Return: Which Course for America's Waterways?" *The Washington Post,* January 10, 2000.

Grusky, Sara. "IMF Forces Water Privatization on Poor Countries." Globalization Challenge Initiative, Institute for Global Communications, February 7, 2001.

Hall, David. "Privatization Often Leads to Corruption." *Sowetan* (South Africa), October 14, 1999.

"Historical Flow Rates in the Everglades: Implications for Management." Southeast Environmental Research Center, Florida International University.

"Infrastructure: The Commission on Building for the 21st Century." State of California. Lieutenant Governor Cruz M. Bustamante, Commission Co-Chair.

Lifsher, Marc. "How Cadiz Inc.'s CEO Became a Key Davis Adviser on Water." *The Wall Street Journal,* December 9, 1998.

"Major New Report Confirms Social, Economic, Environmental Harm from Dams." International Committee on Dams, Rivers and People, November 16, 2000.

McDermott, Terry. "Knee-Deep Disputes for 'Water Buffaloes.'" *Los Angeles Times,* November 1, 1998.

Meltzer, Allan H. "What's Wrong with the IMF? What Would Be Better?" Presentation at "Asia: An Analysis of Financial Crisis," Federal Reserve Bank of Chicago, October 8–10, 1998.

"Ocean Facts: Human Impacts." Pew Oceans Commission.

Perry, Charles A. "Significant Floods in the United States During the 20th Century." United States Geological Survey.

Radhika, V. "I Went to Narmada and Became a Mad Molecule Floating Around." *The Week* (India), August 1, 1999.

Sawyer, Kathy. "Scientists Spot Wildly Spinning Asteroid with Water-Rich Minerals." *The Washington Post,* August 27, 1999.

Shultz, Jim. "Bechtel Says It's Staying." Pacific News Service, April 11, 2000.

———. "Behind Globalization, an Old Demand, Democracy." Pacific News Service, May 4, 2000.

———. "Bloodshed Under Bolivian Martial Law." Pacific News Service, April 9, 2000.

———. "Bolivia Under Martial Law." Pacific News Service, April 8, 2000.

———. "Bolivian Protesters Win War over Water." Pacific News Service, April 7, 2000.

———. "In the Andes Echoes of Seattle." Pacific News Service, March 23, 2000.

———. "Protests Continue As Sides Seek Agreement." Pacific News Service, April 10, 2000.

———. "A War over Water." Pacific News Service, February 4, 2000.

Smith, Michael. "Chile, Bolivia Revive Century-Old Spat over Water." Bloomberg News, June 29, 2000.

Smith, Rebecca, and Aaron Lucchetti. "Rebecca Mark's Exit Leaves Azurix Treading Deep Water." *The Wall Street Journal,* August 28, 2000.

Stevens, William K. "How Much Is Nature Worth? For You, $33 Trillion." *The New York Times,* May 20, 1997.

————. "Water: Pushing the Limits of an Irreplaceable Resource." *The New York Times,* December 8, 1998.

Tomsho, Robert. "Dirt Poor: Colorado Farmers Find Their Water Is Worth More Than Their Crops." *The Wall Street Journal,* April 25, 2000.

Tucker, Katheryn Hayes. "Saying Goodbye to the 'Burbs." *The New York Times,* March 5, 2000.

Wartzman, Rick. "Liquid Assets: A Water-Policy Critic Tries Going Corporate to Tap New Market." *The Wall Street Journal,* March 23, 1999.

"Water to Feed the World: Perspectives for the Future." Food and Agriculture Organization, United Nations.

"William Mulholland and the Collapse of St. Francis Dam." www.usc.edu.

"World Commission on Dams Report Vindicates Unjustifiability of Large Dams." *Dams and Development: A New Framework for Decision-Making.* World Commission on Dams, November 16, 2000.

Yaron, Gil. "The Final Frontier: A Working Paper on the Big 10 Global Water Corporations and the Privitization and Corporatization of the World's Last Public Resource." Citizens' Council on Corporate Issues, March 15, 2000.

Zachary, G. Pascal. "Narmada: World Bank Forces Battle of the Dams to the Table." *The Wall Street Journal,* March 19, 1998.

WEBSITES

www.americanrivers.org

American Rivers Organization, an environmental watchdog group that focuses on protecting U.S. rivers.

www.awra.org

American Water Resources Association, a group offering an information exchange on development of water resources.

Notes

www.awwa.org
American Water Works Association, an international scientific and educational society for water industry professionals devoted to improving the quality and supply of drinking water.

www.canadians.org
Council of Canadians, anti-globalization activists.

crunch.tec.army.mil/nid/webpages/nid.html
An inventory of dams in the United States, as required under the 1972 National Dam Inspection Act.

www.dams.org
World Commission on Dams, which represents all sides of the debate in reviewing the effectiveness of large dams and assessing alternatives.

www.edf.org
Environmental Defense Fund, an advocacy organization.

www.gwpforum.org
Global Water Partnership, an activist organization that helps countries develop plans for sustainable management of water resources.

www.ifg.org
International Forum on Globalization, an alliance of anti-globalization activists, scholars, economists, researchers, and writers.

www.irn.org
International Rivers Network, an environmental organization that works with local communities to protect rivers and watersheds.

mars.jpl.nasa.gov/msp98/index.html
Mars Polar Lander, a site dedicated to studies on the search for water on Mars.

www.msss.com
Malin Space Science Systems, a company that claims to have discovered water on Mars through photographs taken from its Mars Orbiter Camera.

www.narmada.org
Friends of River Narmada, a group opposed to the building of the Sardar
Sarovar Dam in India.

www.nilebasin.com
Nile Basin Society, an educational group created to increase awareness of
the Nile River water crisis.

wrri.nmsu.edu/niwr
National Institutes for Water Resources, a network of water research
institutes in the United States.

www.oceansonline.com
Remarkable Ocean World, an educational site concentrating on ocean
issues.

www.pacinst.org
Pacific Institute for Studies in Development, Environment and Security,
which conducts research and policy analysis on the environment, sustainable development, and international security.

www.polarisinstitute.org
Polaris Institute, an anti-globalization organization that trains citizen
groups to oppose privatization and corporate involvement in local affairs.

www.reefrelief.org
Reef Relief, an environmental group that works for the protection of
coral reefs.

www.seaweb.org
SeaWeb, an educational site that disseminates information about the
world's oceans and the life within them.

www.sierraclub.org
Sierra Club, an environmental organization.

www.turnpoint.org
Turning Point Project, which produces educational advertisements con-
cerning water, the environment, and globalization.

www.un.org/overview/rights.html
United Nations site featuring the Universal Declaration of Human Rights.

www.uswaternews.com
U.S. Water News Online, a service that provides breaking stories about water.

www.waterinvestments.com
A site offering business and finance information for the water industry.

www.waterrightsmarket.com
Water Rights Market, an online marketplace for buyers and sellers of water, water rights, and water-related properties in the western United States.

www.worldbank.org
World Bank, an international agency that provides funding for water projects.

www.worldwater.org
World's Water, a clearinghouse for information on global freshwater problems and solutions.

www.worldwaterforum.org
World Water Forum, a World Bank–sponsored meeting on global water issues.

www.wsta.org.bh
Water Science and Technology Association, a research organization that focuses on water science in the Arabian Gulf region.

Index

Aegean, Sea, 119
Agency for Intellectual Development, 10
agriculture, 2–3, 106
   on Nile riverbed, 27–28
   irrigation, 8, 17, 70, 140, 146, 166
      contracts up for auction, 101–2
   sugar farmers, 147–49
Aguas de Tunari, 108
Alaska, 126
Albright, Madeleine, 153–54, 156–57,
   172
Amazon River, 5
Aquapolaris, 120
Aquarius Water Trading and Transportation,
   121–24
Aqueducts, 62
Aquifers, 127, 135
   Biscayne, 148–49
   Flint River, 2–3
   in Florida, 142
   Ogallala, 166
Aswan High Dam, 9, 74
   creation of, 28–29
   diplomatic rift over, 52
Atlanta
   population growth, 2

Atlantis, 179–81
Azurix, 67

Babbitt, Bruce, 68
Bag shipments of water, 121–26, 137
   debate over, in Canada, 128
   in Piraeus, Greece, 119–22
Baptism, 188–89
Barlow, Maude, 127–28, 132
Bechtel corporation, 12, 107–14
   Augas de Tunari, 108
Bilharziasis, 83
Biscayne Aquifer, 148–49
Black March, 91
Blair, Tony, 116
"Blue Gold" (Barlow), 127
bottled water, 4, 135

Cabora Bassa Dam, 22
Cadiz Groundwater Storage and Dry Year
      Supply Program, 47, 48, 50, 65
Cadiz Lake, 47–50, 65–66
California, 3, 55–56
   population growth, 63
Canada
   bulk removal of water, 132

Canada (*cont.*)
  compared to Saudi Arabia, 128
  transfer of water, 129, 133
  rejected water projects, 129
  resistance to selling water, 131
Canyon Lake Dam
  flood, 16, 30–41
  Missouri River, 15
  Rapid City, 30–38
Chattahoochie River
  dispute over control of, 1–2
Cholera, 5, 83, 100
Clearwater project, 190–91
Cochabamba, Bolivia, 107–17, 134–35
  Permanent Assembly on Human Rights, 111
  riots in, 12, 99
Colorado River, 25, 68–69
  Hoover Dam, 23–24
Companies, water-related
  Aquapolaris, 120
  Aquarius Water Trading and Transporta-
    tion, 121–24
  Bechtel corporation, 12, 107–14
    Aguas de Tunari, 108
  Suez Lyonnaise des Eaux, 11, 101–4
  Vivendi SA, 11
  Wessex Water, 101
Conservation, Mandatory, 164
Controlling water flow, 17.
  automization, 146–47
  See also Dams.
Coral reefs, 177–78
Council of Canadians, 127

Dams, 126
  Aswan High Dam, 9, 74
    creation of, 28–29
    diplomatic rift over, 52
  Cabora Bassa, 22
  Canyon Lake, 15, 16, 30–41
  consequences of, 17
  Edwards, 142
  Hoover
    history of, 23–24
    as tourist attraction, 26
  hydroelectric, 18, 19, 93
  Oahe, 37
  removal of, 144, 172
  Sardar Sarovar, 90–93
  subsurface, 83
  Xiaolangdi, 93
  See also specific rivers and locations.

Deals, water-related, 101
Defense of Water and Life, 110
Delivery of water, 103, 105, 106
Democracy Center, 115
Desalination, 174–78, 190
  costs of, 177
  financial support for, 175
  in Kuwait, 155
  in Tampa Bay, 176
  techniques, 175
Desert Center, California, 43–45
Diseases and bacteria, water-related
  bilharziasis, 83
  cholera, 5, 83, 100
  dysentery, 5
  E. coli, 102
  malaria, 152
Diversion of water bodies, 126, 173. *See also*
    Dams, *specific bodies of water.*
Droughts
  Florida, 144
Dust
  respiratory illnesses, 108
Dysentery, 5

Earth Day, 153, 156–57, 172
E. coli, 102
Edwards Dam, 142
Egypt
  ancient, 26
  Aswan High Dam, 9, 28–29, 52, 74
  Environmental Affairs Agency, 75
Electronic water auctions, 101–2
Endangered species, 38, 169
  in the Everlades, 141–42
  Pacific salmon, 167–72
England
  approach to water management, 116–17
Environmental problems, 64, 65, 136
  as consequence of water projects, 152
  destruction of fish habitats, 152
  as effect of diversion of river, 58–60
  in Los Angeles, 63
Environmental Protection Agency, 105
Everglades, 139, 144, 147
  endangered species in, 141–42
  habitat, 141
  restoration of, 140, 142, 149–50
Extinction, 127

Fanjul family, 147–48
Federal Bureau of Reclamation, 61

Federal Emergency Management Administration (FEMA), 41
First Nations coalition, 128
Fish
  destruction of habitats, 152
  overfishing, 169
  salmon, 167–72
Flint River and Aquifer, 2–3
Floods, 20–23
  as consequence of dams, 20
  control projects, 17–18
  environmental effects of, 60
  in rainy seasons, 21
  See also specific rivers and locations.
Florida
  Everglades, 139, 141–42, 144, 147, 149–50
  population growth, 145
  sugar farmers, 147–49

Gambia River, 21
Ganges River, 9
Glaciers, 120, 128
Gleick, Peter, 4–6, 13–14
Globalization, 12, 99–101, 113–14
  demonstrations against, 134
Gorbachev, Mikhail, 86, 160–61
Great Lakes, 127, 136
Great Stalin Plan for the Transformation of Nature, 19
Green Cross International, 161–63
  compact, 163

Hoover Dam
  history of, 23–24
  Franklin D. Roosevelt and, 24
  as tourist attraction, 26
Hydroelectric
  dams, 18, 19, 93
  plants, 19
Hydrothermal vents, 179–81

Indanglasia
  gender roles, 80, 83–84
  pump fund, 82–83
  Nzoyia River, 80
  theft of pumps, 84–85
  transporting water, 81
Industry
  corporate development, 164
  water needs of, 8, 106, 166, 173
International Conference on Water and the Environment, 88, 115, 131

International Joint Commission, 136
International Monetary Fund, 115
Iraq, 155
Irrigation, 8, 17, 24, 70, 140, 146, 166, 173
  contracts up for auction, 101–2
  water management for, 17
Islam, 53, 96–97

Jordan River, 52, 154–55, 189

Kairakkhum Channel, 160
Kennebec River, 142–43
Kuwait, 155

Lakes
  Cadiz, 47–50, 65–66
  Chad, 9
  Gisborne, 133
  Great, 127, 136
  Mead, 25
  Nasser, 28, 74
  Oahe, 37
  Okeechobee, 138–41, 145, 147, 148
  Tahoe, 58
Lesotho, South Africa, 150–51
Lewis and Clark, 36
Liters per day of water use, 4–5, 7, 92, 94, 100
Lockhart, Teena and Jim, 38–40
Los Angeles, 3
  climate, 55–56
  population growth, 63
Los Angeles River, 56

McCurdy Enterprises, 133
Malaria, 152
Mao Zedong, 19
Managing water, 93, 164
  justification of, 17
Mars
  Polar Lander, 182–85
  water on, 173, 182–85
Massachusetts
  water management, 105
Massachusetts Water Resource Authority, 103
Metropolitan Water District, 66–68, 70, 71, 124
Mid-Atlantic Ridge, 179
Middle East, 156
  lack of water, 51
Mississippi River
  flood, 41
Missouri River, 15
Mount Whitney, 59–60

Nasser, Gamal Abdel, 28
New England Water Environment Associa-
    tion, 104
Nile River, 51, 74
    dam, 18
    erosion of riverbed, 29
    yearly flood cycle, 27–28
Nongovernmental organizations (NGOs), 86–
    87, 89
North American Free Trade Agreement
    (NAFTA), 131–33
North American Water Power Alliance
    (NAWAPA), 130–31
Nzoyia River, 80

Office of Water Services, 116
Ogallala Aquifer, 166
Okavango River, 21, 158–59
Olivera, Oscar, 134
Owens River, 57–65

Paraguay-Paraná Hidrovia project, 150
Persian Gulf War, 155–56
Pilcomayo River, 162–63, 172
Pipelines, water, 3, 105–6, 123
Political conflicts over water, 154, 160, 164–
    65
    Aswan High Dam, 28–29, 52
    Botswana and Nambia, 10, 158
    Chatahoochie River, 1–2
    Middle East, 10
    Singapore and Malaysia, 157
    Texas, 166–67
    Turkey, 10
    Virginia and Maryland, 165
    See also Wars over water.
Political Economy Research Center, 96
Pollution, 103, 120, 125, 127, 162, 173
    as consequence of dams, 20
    toxic metals, 59
Population growth, 164
    Atlanta, 2
    California, Nevada, and Arizona, 75
    Florida, 145
    Los Angeles, 63
Potomac River, 165
Pricing of water, 96, 106–7, 108–9, 124
Privatization of water companies, 12, 103,
    115–17
    environmentalists' stance on, 127–28
    protests over, 110
    as solution to water crisis, 190

Protests
    against globalization, 134
    over water, 110
Public water companies, 106
Puget Sound, development of, 168–69
Purification plants, 3

Rabin, Yitzhak, 54
Rainfall, 8, 52
Regulation of water, 116
Reservoirs, 18
    subterranian, 84
Respiratory illnesses, 108
Restoring waterways, 143–44
Reverse osmosis, 175
Right vs. need, water as, 78–79, 87, 95
Rio Earth Summit, 88, 131
Rio Grande, 16
Rivers
    Amazon, 5
    Colorado, 23–24, 25, 68
    Ganges, 9
    Chattahoochie, 1–2
    Gambia, 21
    Jordan, 52, 154–55, 189
    Kennebec, 142–43
    Los Angeles, 56
    Mississippi, 41
    Missouri, 15
    Nile, 18, 27–28, 51, 74
    Nzoyia, 80
    Okavango, 21, 158–59
    Owens, 57–65
    Pilcomayo, 162–63, 172
    Potomac, 165
    Rocha, 107
    Sacramento, 124
    Skagit, 170–72
    Snake, 167–68
    Tennessee, 19
    Tigris and Euphrates, 18, 51, 53–54
    Volga, 19
    Yellow, 20, 93
Rocha River, 107
Roosevelt, Franklin D., 19, 24
Roosevelt, Theodore, 60–61

Sacramento River, 124
Salmon, Pacific
    as endangered species, 167–72
    overfishing, 169
    spawning, 168

San Francisquito Canyon, 62
Sardar Sarovar Dam, 90–93
  Black March, 91
  World Bank and, 90–93
Sewer systems, 105–6
Sierra Club, 71, 89
  California/Nevada Desert Committee,
    73
Six Day War of June 1967, 51–52
Skagit River, 170–72
Skagit Watershed Council, 171
Snake River, 167–68
Southwestern U.S. population growth, 75
Sprawl, urban, 164
Stalin, Josef, 19
Subsurface dam, 83
Subterranian reservoirs, 84
Suez Canal, 29
Suez Lyonnaise des Eaux, 11, 101–4
Supply and demand, 108
Swedish Landscape Ecology Group, 20

Tampa Bay desalination project, 176
Tennessee River, 19
Thatcher, Margaret, 116
Thermal Distillation, 175
Three Gorges Dam, 151
Tigris and Euphrates, 18, 51, 53–54
Transfer of water, 129, 133, 135, 136,
    190
  bag shipments, 121–26, 137
    debate over, in Canada, 128
    in Piraeus, Greece, 119–122
  and Canada, 129, 133
  by foot, 81
  pipelines, 3, 123
Turkey, 53

Union Pacific Railroad, 55
United Nations, 82
  Declaration of Human Rights, 78–79,
    93
  subsidization of water imports, 125
U.S. Geological Survey, 72
United Water Resources, 101
Unsanitary conditions, 95
Utilities, water, 102

Vietnam
  access to water, 3
Violent acts over water, 159–60
"Vision" report, 87–90

Vivendi SA, 11
Volga River, 19

Wars over water, 10
  Six Day War of June 1967, 51–52
  See also Political conflicts over water;
    Violent acts over water.
Water
  as commodity, 87
  as critical element of life, 7, 104–5
  as good vs. resource, 131
  delivery of, 103, 105, 106
  management of, 17, 93, 164
  mismanagement of, 120, 126
  pricing of, 96, 106–7, 108–9, 124
  privatization of companies, 12, 103, 115–
    17
    environmentalists' stance on, 127–28
    protests over, 110
    as solution to water crisis, 190
  purification plants, 3
  quotas, 161
  regulation of, 103
  as right vs. need, 78–79, 87, 95
  scarcity of, 94
  systems, urban, inadequate, 105
  transfer of, 129, 133, 135, 136, 190
    bag shipments, 119–26, 137
    and Canada, 128, 133
    by foot, 81
    pipelines, 3, 123
  usage
    liters per day, 4–5, 92, 94, 100
    utilities, 102
Water bags. See Transfer of water.
Water-relief organizations, 89. See also
    specific organizations.
Water table, 8
Wessex Water, 101
World Bank, 4, 79, 89, 117, 131, 173
  and Sardar Sarovar Dam, 90–93
  subsidization of water imports, 125
  water-management record, 90
World Health Organization, 4
World Trade Organization (WTO),
    131
World Water Council, 86
World Water Forum, 86, 88–89

Xiaolangdi Dam, 93

Yellow River, 20, 93

About the Author

A former *Business Week* editor and national editor at Bloomberg News, Jeffrey Rothfeder is the author of *The People vs. Big Tobacco* and *Privacy for Sale*. He has been reporting on water issues since 1979 for publications such as *The Washington Post* and the *St. Petersburg Times,* and has spoken at Congressional hearings and appeared on national television and radio programs including *Nightline, 20/20, Today, All Things Considered, Larry King Live,* and *The Oprah Winfrey Show*. Rothfeder lives in Bronxville, NY.